Cuyahoga's CHILD

Mist on the river. Sleepy and secluded, the Cuyahoga River (near Riverview Road) often projects an image that belies its powerful influence on the United States, northeastern Ohio, and millions of lives that are or were oblivious to this short, seemingly confused river. *Photo by Ian Adams.*

Cuyahoga's Child

GROWING UP IN THE VALLEY OF THE CROOKED RIVER

Jeffrey J. Knowles

Orange *frazer* Press
Wilmington, Ohio

ISBN 978-1939710-307
Copyright©2015 Jeffrey J. Knowles

No part of this publication may be reproduced in any material form (including photocopying or storing in any medium by electronic means and whether or not transiently or incidentally to some other use of this publication) without the written permission of the copyright holder except in accordance with the provisions of the Copyright, Designs and Patents Act 1988.

Published for the author by:
Orange Frazer Press
P.O. Box 214
Wilmington, OH 45177
Telephone: 937.382.3196 for price and shipping information.
Website: www.orangefrazer.com
Book and cover design: Alyson Rua and Orange Frazer Press

Library of Congress Cataloging-in-Publication Data

Knowles, Jeffrey J., 1948-
  Cuyahoga's child : growing up in the valley of the crooked river / Jeffrey J. Knowles.
     pages cm
  Includes bibliographical references and index.
  ISBN 978-1-939710-30-7 (alk. paper)
1. Knowles, Jeffrey J., 1948---Childhood and youth. 2. Northfield (Ohio)--Social life and customs--20th century. 3. Cuyahoga River Valley (Ohio)--Social life and customs--20th century. 4. Cuyahoga River Valley (Ohio)--History. I. Title.
  F499.N698K56 2015
  977.1'31--dc23
                        2015017006

Printed in China

Other services and support for this book were also provided via the courtesy of:

The Kent State University Press, for access to Jack Gieck's *A Photo Album of Ohio's Canal Era, 1825—1913*, and Frank Wilcox's *Ohio Indian Trails*
Ian Adams, of Ian Adams Photography
Tom Jones, of Buckeye Photos
The photography of Q.T. Luong; QT Luong/terragalleria.com
The Cuyahoga Valley National Park
The Akron-Summit County Public Library
The Nordonia Hills Branch of The Akron-Summit County Public Library
The Summit County Historical Society of Akron, Ohio
The Canal Society of Ohio

Dedicated to my family,

Marie, Lorence and Lloyd Knowles,

And to the "FSKs", Frank, Doris, Janet, David, and Jack

And to the memory of the Cuyahoga Valley's

great caretaker, Joe Jesensky

Acknowledgments

There is some risk in trying to acknowledge the contributions to a project over a quarter of a century old. Many helpful people from northeastern Ohio libraries, historical societies, colleges and universities, numerous local governmental and civic agencies, and ordinary people who were and are passionate about their homeland in the valley of the Cuyahoga all left their touches in this book. To these I can only offer my spiritual rather than physical gratitude.

Others I remember who made significant and/or recent contributions deserve something better. The late Joe Jesensky, my Cuyahoga Valley godfather, provided several years of encouragement and personal investment in the forms of letters, visits, and a couple of his signature traipses through the valley, especially the section near our homestead. Steve Paschen, former director of the Summit County Historical Society (SCHS) and Kent State professor emeritus, has been consistently supportive over a long number of years, right down to his contribution of the "Foreword" to this work. Others who offered inspiration along the way include another former SCHS director, Dan Rice (now President and CEO of the Ohio & Erie Canalway Coalition), and Julia Morton, former senior editor at the KSU Press. I also drew encouragement from

William D. Ellis, author of *The Cuyahoga* in the Rivers of America Series, whose book (and a later visit) first struck a meaningful nostalgic chord in my reconsideration of my childhood.

Recently, several people have given me a great deal of help in getting this book out of the door, among these: Mary Plazo, Special Collections Division Librarian of the Akron-Summit County Public Library; Mary Pat Doorley, Office of Public Affairs, Cuyahoga Valley National Park; Carol Heller, Administrative Assistant, KSU Press; and Leianne Neff Heppner, President and CEO, Summit County Historical Society. I also want to acknowledge Ian Adams, environmental photographer extraordinaire, who went beyond the call of duty in his personal interest in, and support for, this venture, and artist Kristin Calhoun, who provided the beautiful homestead and Indian trails maps.

Finally, I am especially grateful for the hours invested by several good people in editing this work. Virginia Case Bloetscher, whose 1987 book, *Indians of the Cuyahoga Valley and Vicinity*, gave me crucial direction for the chapter on the Indian nations in the valley, also contributed much time to editing that chapter and correcting several of my early misconceptions, as well as offering numerous rich insights. Then, there were the superb efforts from family members. Dr. (and cousin) Jack Knowles, professor of English, provided an exhaustive, line-by-line edit of the entire work several years ago when the work was still in draft stages. More recently, my son, Mark, edited the entire work with an eye not only for detail, but also for content. Many changes were made—mostly by subtraction—in response to his very good reader-relevant insights. Finally, my wife of forty-six years, Lezlee, made a huge time investment in reading and rereading chapters these past several months, keeping them corrected and straight at a time when I was drowning in drafts.

At end, I would like to also thank Marcy and Sarah Hawley of Orange Frazer Press for their warm and patient support for this book. It was only when I began working with them that the thing started feeling like a real book.

—JK

Contents

page xv
Foreword by Stephen Paschen

page xviii
Author's Note

page xx
Prelude

---

1. *page 2*
   Feet of Clay
   The Land

2. *page 36*
   Valley Life Line
   The Canal

3. *page 70*
   Altered Altars
   The Buildings

4. *page 118*
   Straight Lines in Curved Space
   The Railroads

5. *page 144*
   Owners of the Clouds
   The Indians

6. *page 186*
   Life Down Under
   The Living Things

7. *page 228*
   An Odd Brew
   The Reserve

8. *page 272*
   Which Way is Up?
   The Faith

page 326
Epilogue

---

page 330
About the Author

page 332
Bibliography

page 338
Index

Foreword
by Stephen Paschen

Memory is a powerful, compelling part of human consciousness. Our brains observe and interpret everything we experience, constantly changing and reconstructing what we think happens and why. Human memory, of course, is flawed. Age, disease, and other influences such as books, popular culture, news media, and even repeated retelling of the same story (our stories evolve as we retell them) affect accuracy. Also, a story told to others is often given a different spin or emphasis when it is passed on and details and facts become muddled. However, despite its drawbacks, memory is a useful contributor to the preservation and interpretation of history. The fields of public history (the history of everyday people) and oral history (the recorded testimony of human experiences) accept memory as a viable source. Memory captures perspective and provides evidence of human emotions, causal forces in everyday life as well as historical events.

In this book Jeff Knowles uses his own personal memories along with traditional historical sources to link his life to regional history. His is a public historian's approach to understanding the "Baby Boomer" generation (of which he is a member) who have lived in the Cuyahoga River valley, part of Ohio's Western Reserve. The interplay between natural and human history related in his own distinctive voice is a very effective way of connecting the reader to the recent and distant past. It's a little like exploring the land and its history with Knowles along as your guide. Historical research, storytelling, memory, deductive reasoning and effective historical context are melded into a cogent, readable book.

Knowles's personal approach resonates with me, perhaps because during my earliest years living in Ohio (I moved here in the spring of 1973) my profession at that time was landscape architecture. In my dozen years as a landscape architect (I later became a public historian and archivist) I worked in both private and public practice doing landscape design and urban planning. Any new landscape design or architectural project was preceded by researching the past as well as the existing natural and built environment within the project area. This is precisely Knowles's approach to this history. He brings to his work a keen lifelong curiosity and imagination about the landscape in his own backyard and extended surroundings, which enables him to postulate connections between recorded history and observed physical traces left upon the land.

Knowles uses an unusual vehicle (for a historian) in this history. He creates fictitious but believable characters (such as his plausible nineteenth century homestead predecessor, the boy Samuel) to give voice to the people who passed through the region over time. His musings about Samuel provide a different perspective for considering historic peoples in the context of the author's own life. He recounts the arrival and influence of white settlers, whose industriousness, cultures, and mores made their imprint on history and influenced modern life in the Western Reserve.

Knowles is observant of visible physical traces of the past, including marks left by Ohio's canals, architecture, railroads, natural and human-made changes in the lay of the land, political and property boundaries (including the origins of the Western Reserve), and relates religious influences brought here by diverse residents. Images, illustrations and maps provide orientation and visual cues to his interpretation of the region's history. The story is not told chronologically. Instead it is a many-faceted mixture of origins, memories, impressions and soundly researched history providing context for modern life and the cultural, spiritual and physical phenomena that shaped the Western Reserve.

Whether one is a historian or non-historian, each of us forms our own impressions of what happened over time based on many sources and experiences. Jeff Knowles offers his own impressions based on solid research of the people and the land of the Western Reserve in a fascinating and connecting way. His contemplative journey encourages readers to search their recollections and perhaps conduct a little research of their own. The beautiful place we know as the Western Reserve has inspired countless peoples to interact with the land in different ways that contribute to a rich history. This book reminds us that the stories of everyday lives are vital to the understanding of our past.

—Stephen H. Paschen
Associate Professor Emeritus
Kent State University

Author's Note

Two persistent challenges have attended the writing of this book, one geographical, the other chronological. The geographical challenge relates to old Northfield Township, created in 1797 as Town 5, Range 11 of the Connecticut Western Reserve, and first "settled" by English-speaking people in 1807. It was, like the other some-two hundred townships in the Reserve, roughly twenty-five square miles in size. But it did not stay that way, later dividing into four separate units of government sharing a common school district. The smaller two, Northfield Center and Northfield Village, became, respectively, a township and a village. The larger two, Macedonia and Sagamore Hills, became, in time, a city and a township in Ohio's complicated jurisdictional schemes of local government. In referencing my childhood homestead, I have usually used the term "Northfield" to refer to the larger, original twenty-five square mile area, even though we technically lived in Sagamore Hills. This seems oddly clumsy, but it reflects the order which most of us practiced—Northfield, first, Sagamore Hills, second—in telling people where we lived. I have on a couple of occasions reiterated these distinctions in other places in the book.

The second challenge, far more difficult, stems from the three different time standpoints I struggled to negotiate in this work. As a retrospective, and not an autobiography, I have attempted more than a mere memory project here. I also tried to put those memories into some kind of meaningful context from a later life perspective. The issue is that this later life perspective, itself, has two dimensions: my original drafting era when I was in my early to mid-forties, and my publication timeframe right now, in my mid-sixties. Any author will tell you that he or she cannot leave an unpublished manuscript alone (i.e., without more edits when rereading). There is always one more edit; only the steely throat of a printing machine can turn the endlessly chewed manuscript into a fully digested book. Hence, a good reader will see that my later/later life perspective inevitably creeps into this book. I admit the ambiguity, but have worked to avoid it wherever possible. As a general rule, I have tried to maintain the mid-life context of the original drafting.

One other note, this one a reminder of sorts: Memory, while a crucial resource for history, does not—cannot—meet all of history's standards for accuracy. Nor is that intended here. Indeed, it is often memory's slanted stage that provides historians with valuable insights into the past. Some of my memories herein carry the distortions inherent in the mind of an eight or nine-year old. This does not give me a free pass on historical accuracy—no one claiming to write history can be permitted such a thing—only, perhaps, a hope for some indulgence on the reader's part.

Prelude

Kendal Lake reflection. CVNP. Sometimes the past, behind or beneath us, gives a clearer picture of reality than the present. Not a bad place to begin for people attempting to understand their lives—and life. *Photo by Q.T. Luong.*

There is a stretch of land in northeastern Ohio that is, to the casual observer, entirely ordinary. It is a cup-shaped piece of earth anchored on the top and bottom by large cities and carved by a confused river that at times is no bigger than a healthy creek. The couple of million people who live inside the cup seem not too different from other splashes of once-industrialized humanity clustered in Detroit or Youngstown as they curse their way through pot-holed roads in the cold of winter. It might be taken for just another endless spread of malls, condos, and fast food huts set against the backdrop of aging empty factories such as those that often dot chunks of American real estate east of the Big River, except…

…except for the fact that this piece of land is, arguably, the historical heart of the nation. Observers such as George Washington and Thomas Jefferson saw the outline of its destiny immediately, long before there were any malls or factories—or white-skinned settlers, for that matter. The continent's watershed thins to a narrow high ground at modern Akron, sending cascading rivers north to Lake Erie and south to the Ohio River. The bit of dry ground topping this hog's back held the secret to making the United States into an economic and physical whole. It is a space charged with electricity, like the tiny gap separating the fingers of God and his lethargic man of clay in Michelangelo's *Creation of Adam*, only here the reaching arms are the huge St. Lawrence/Great Lakes seaway stretching southwestward from the dynamic father waters of the Atlantic, and the often sleepy Mississippi/Ohio/Muskingum/Tuscarawas River chain stretching up from the smaller Gulf of Mexico. Bordering the few miles of high ground spaced between the ends of those watery fingertips lies the southern loop of the crooked U-cup sculpted by the meandering Cuyahoga River.

There was a boy who grew up just inside the western wall of the cup, halfway between Akron and Cleveland. In many ways his 1950s childhood was uneventful, not unlike the river which then, except at its mouth in Cleveland, was largely ignored as it flowed north to its dirty industrial destination. The boy lived on a plot of ground that had been trampled by some of the greatest military figures in the history of the continent: Iroquois and Ottawas and a wandering Shawnee chief, Indians who held the land for what is still more than half of the white man's time in the "new world." The boy's home overlooked the magnificent river valley (were the thick trees to have permitted such a look), and at the base of the valley wall, the foot of his hill, stood the overgrown ruins of a lock on the Ohio & Erie Canal. A few feet west of Old Red Lock flowed the waters that once marked the western boundary of the United States. But the boy knew none of this. He grew up picking blackberries and playing baseball and traipsing through heavy snows to catch the school bus, all the while assuming that his life was entirely ordinary, just like the common dirt under his feet and home.

This is the story of two childhoods, which were both ordinary only to the extent that God's extraordinary gifts of life and living can be called ordinary.

The "childhood" of the Cuyahoga River assumes some liberties, even in the world of metaphor. The Cuyahoga, old by most human standards, is still in its archaeological infancy, having gotten its latest trimming from the geologically recent Wisconsin Glacier. From the myopic viewpoint of the white settlers, the river lived the best years of its youth during the first half of the nineteenth century. There, for a historical moment, it was the darling child of an adoring people. Yet, never a very practical child. Beginning in the middle of Geauga County, just a few miles east of Cleveland and actually north of the point where it eventually boomerangs back to that city, the Cuyahoga spurns the shortest distance to its mouth by heading south where it vainly bounces up against the high ridge of America's watershed at what is now Akron. Thus rebuked, it turns back toward its ultimate outlet at Cleveland on Lake Erie, covering a distance nearly five times that of a straight line between its headwaters and mouth. Along the way it tossed up major American cities on its slim banks and floated the nation's heavy steel industry on its hairpin curves.

My own childhood story was more timidly scripted. There are no movie rights or TV pilots pulsing within it. Few of the main actors get beaten up, or shot up, or high on drugs, or even divorced. We had a dog that was not killed while defending our lives, and a church in which the minister did not carry on with the organist. Thus insulated, it was not until I was middle-aged that I realized my unremarkable existence might, after all, be rather remarkable, or at least worth a comment. There are millions of others out there who, like me, are in the same boat. Weaned on Campfire Girls, Little League and Rock 'n Roll we spilled out of two-parent homes into a world that would later have us feel at best boring and at worst guilty for being ordinary. But in a world where the extraordinary has become ordinary, the ordinary begins to look extraordinary. There are things worth remembering about how we grew up. And where.

Yet, this book is not a golden-hued remembrance of that which never was, and which, by its very distance, is now safe for the telling. Sentiment is a healthy human emotion only when rooted in informed honesty. In point of fact, the 1950s of my childhood was not, in many ways a pleasant time, just as the early 1820s of the Cuyahoga's "youth" was not a very pleasant time. By the time we Baby-Boomers came calling, the world's most ruinous half-century had robbed people of much of their moral certainty. America was no longer the fair-haired wonder of the international community. Her dried-up churches were struggling to retain tiny slivers of light from the religious quasars that had brought them into existence in the eighteenth and nineteenth centuries. The technological explosion with its corollary stream of "conveniences" masqueraded for a time as a means rather than an end, but even in the '50s the charade was wearing thin. Too, there was a perpetually oozing tension that smothered the land as an increasingly anxiety-ridden people worried about the month left at the end of their money, polio, The Bomb, and why it was so much easier to hate than to love. There were no more "good old days" during the 1950s than there are now.

But for many decades, far beneath the headlines, God's consummate building blocks—human stories acted out on the stage of his second-most prized gift, life—were quietly fashioning temples of great value. Tiny people like my Grandma Hendricks were desperately clinging to the threads of broken lives long after others had let go and fallen into familial oblivion. Somehow some things of value survived, a memory here, a tradition there, a promise tucked away but never abandoned.[1] Tiny people were struggling in a most unglamorous way to brake the downward slide, to gain a foothold, setting the stage for their children, my parents' generation, to make their remarkable turnarounds. There had to be millions of those people who contributed their life stories to make ours possible—people who got out of bed to go to work years after that task had ceased giving them any satisfaction or dignity, who held on to marriages because they put at least one other person in the family ahead of themselves, who kept staggering forward under the hammer blows of life long after crude obedience had superseded hopeful dreams. Those stories are seldom told. We seem to prefer talking about our failures.

How do you tell such stories? How do you combine the childhood of a boy with that of a river? I found that I could do no better than to simply and honestly narrate those times as they came to ground in the geographical and personal places of my childhood. For this reason, the chapters in this book are essentially nouns. They speak of the land, buildings, creatures and other physical aspects that filled up my young life on the hilly banks of the Cuyahoga, and that filled up the valley for generations before I was born. But something far more than the physical environment was at work here. Those seeming props on the stage became part of the actors, thereby creating lives that were infinitely diverse and uniquely significant. At the end of the day, *where we were* has a great deal to say about *who we are*.

Ultimately, all that I can conclude about this work—nothing more, but nothing less—is that this passage through the valley of two childhoods is worth remembering—and perhaps even treasuring.

Notes

1. See Roy Rosenzweig and David Thelen, *The Presence of the Past* (New York: Columbia University Press, 1998) for a fine discussion of Americans' intuitive hunger for the past, a thesis that goes against the current assessment of many professional historians, and one that encourages works such as this one.

Cuyahoga River Valley

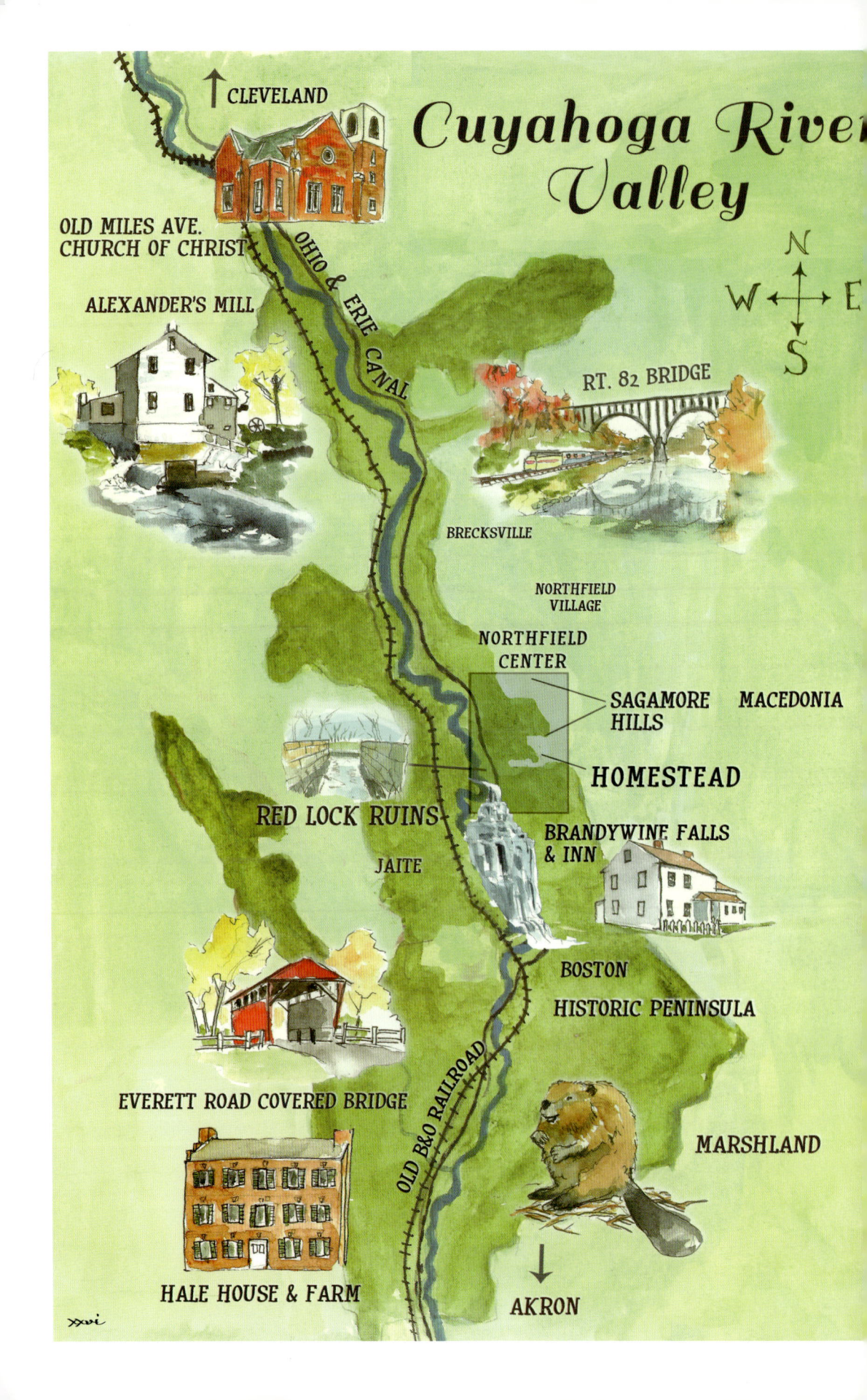

↑ CLEVELAND

OLD MILES AVE. CHURCH OF CHRIST

ALEXANDER'S MILL

OHIO & ERIE CANAL

RT. 82 BRIDGE

BRECKSVILLE

NORTHFIELD VILLAGE

NORTHFIELD CENTER

SAGAMORE HILLS

MACEDONIA

HOMESTEAD

RED LOCK RUINS

JAITE

BRANDYWINE FALLS & INN

BOSTON

HISTORIC PENINSULA

EVERETT ROAD COVERED BRIDGE

OLD B&O RAILROAD

MARSHLAND

HALE HOUSE & FARM

↓ AKRON

xxvi

Enlarged map from shaded area on opposite page

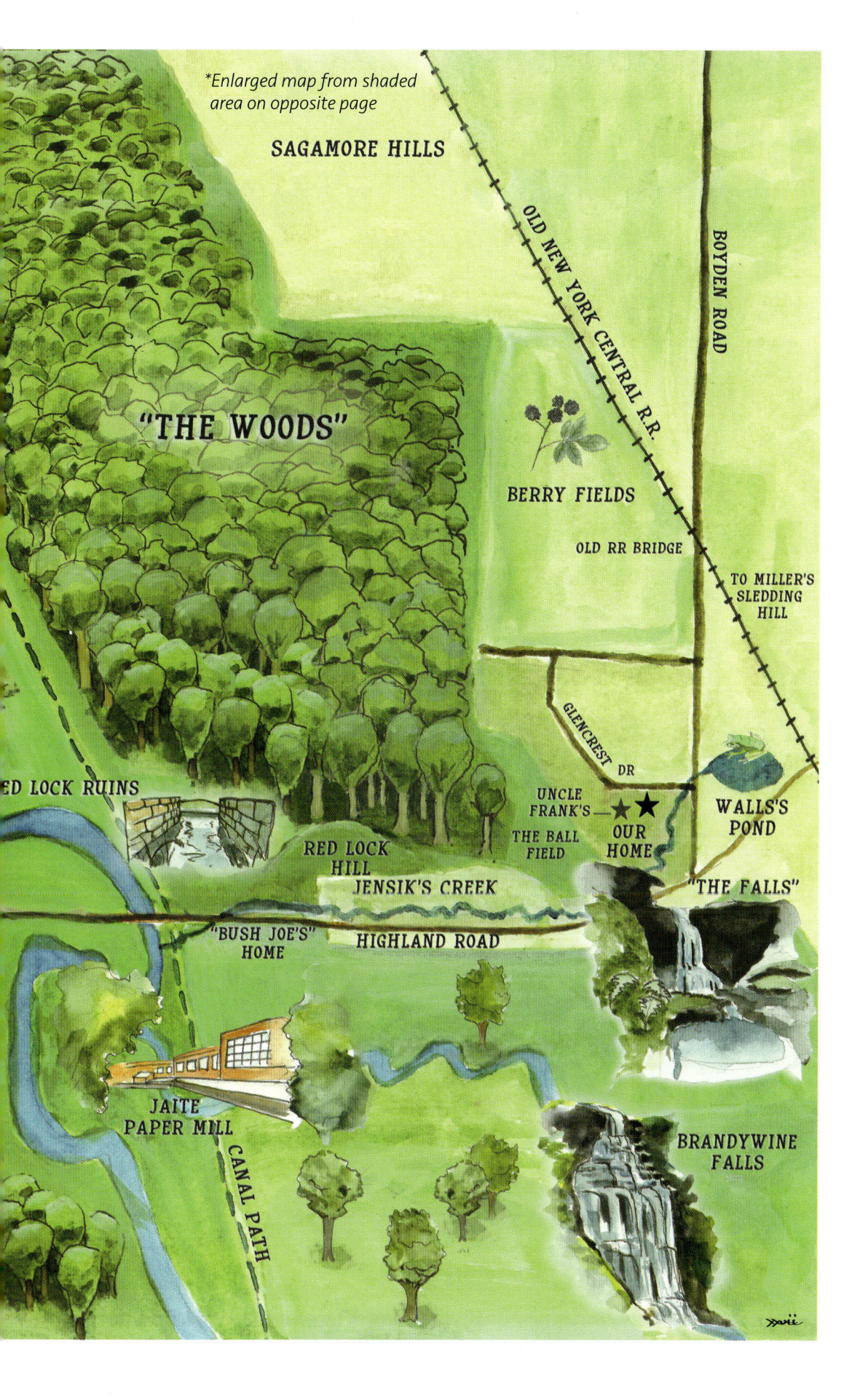

SAGAMORE HILLS

OLD NEW YORK CENTRAL R.R.

BOYDEN ROAD

"THE WOODS"

BERRY FIELDS

OLD RR BRIDGE

TO MILLER'S SLEDDING HILL

GLENCREST DR

RED LOCK RUINS

RED LOCK HILL

JENSIK'S CREEK

UNCLE FRANK'S

THE BALL FIELD

OUR HOME

WALLS'S POND

"THE FALLS"

"BUSH JOE'S" HOME

HIGHLAND ROAD

JAITE PAPER MILL

CANAL PATH

BRANDYWINE FALLS

Cuyahoga's CHILD

1
Feet of Clay
The Land

Brandywine Falls. Perhaps the signature scene in the Cuyahoga valley and a very popular attraction in the Cuyahoga Valley National Park, Brandywine Falls is just over a mile from the old Knowles homestead. *Photo by Ian Adams.*

The sculpting of the Cuyahoga Valley • The gifts of the land • The ball field, falls, and ravine • The best of creation's hand

"Dig your fingers into the ground," my brother, Lloyd, shouted down to me. His voice conveyed an air of authority—he was six while I was a third of a lifetime behind him at four—but also something of the excitement of discovery. As with most of our fraternal wanderings through childhood he had again been the first to uncover the secret, to find the treasure, to learn the lesson.

"It's easy if you dig your fingers into the ground. You can climb right up. Watch me." I would have found his confident assertions irritating had I been anything less than terrified at that moment. I tried digging my fingers into the ground but couldn't get the hang of it. Behind and all around me was the most unnerving and intriguing landscape this city boy had ever seen. Enormous rock outcroppings jutted out over a canyon that to my fevered eyes looked four- or five-hundred feet deep, but was actually only thirty or forty. The canyon ringed the pearl of this Tolkienesque scene, a perfectly formed waterfall that tumbled into a broad, round pool below. Underneath the thrusting rock plates darkening shadows slanted back into what I was sure were deep caves. Below was the creek, splashing and weaving among a disordered array of oddly placed rocks and fallen trees. It was, I realized, a wild world that defied sun, grass stains, mail boxes—and parental protection.

Why had we come down here anyway? Dad and Mom had strictly warned us to stay away from this place. Aunt Doris and Uncle Frank, whose country home we were visiting that day, had spoken of the "falls" in a way suggesting a place that specialized in eating up little children. At the very least I expected to find a troll or two. Yet here we were, halfway up the walls of the steep ravine that framed the dreaded place. And I was stuck: unable to go up, not daring to go down, and not caring to stay put. Lloyd was saying something else from above me. I couldn't hear what he said, but did notice a dramatic change in his formerly bossy tone. Suddenly he was no longer the authoritative older brother, but a scared little boy. Despite my predicament I lifted my nose out of the ground-hugging ferns long enough to steal a look at him. Beyond him a few feet at the top rim of the ravine I saw Dad glowering down at us. What little space I had left in my small body for fear immediately got filled up.

Perhaps it was that extra shot that got me to the top of the ravine, or perhaps I finally made use of Lloyd's advice. I don't remember. I do remember that Dad didn't come down for us, a fact I wondered about. I toyed with the idea that he had determined that a long fall into a rocky stream bottom would be a fitting punishment for our disobedience. Over a half-century later, I realize that the ravine was not the Grand Canyon, and that even a four-year-old's instinctive, spread-eagling capacity for self-arrest would have prevented a long fall. Back up at Uncle Frank's house, Mom had only made it a few feet into the back yard. She was a trained medical professional who would spend her career in psychiatric hospitals helping to heal torn bodies and tortured minds, a person who probably saw more suffering in a month than most people see in a lifetime. But Mom did not react as well to the suffering of her family. She had not accompanied Dad to the rim of the falls for the simple reason that her shaking legs would not take her there. Now she was greatly relieved to see our bobbing heads and hear our pained cries as Dad administered justice all the way back up to the house.

But if danger lurked in that canyon, so too did a magnetic attraction, a calling strong enough to drive two usually obedient boys across the threshold of direct disobedience, a singing siren that promised these city slickers more than a pretty song and a dashing on the rocks. It promised a way of life. It promised to seep inside our veins and to become a part of our marrow, raising up for us the sounds of its broken stream and the sighs of its leaning trees long after we had quit its ragged paths and settled in faraway cities. The call of that land never quite faded from our hearing.

~~~

Topographically, there isn't much that is particularly unique or arresting about the ravine that bumped up against what became our 1953 property line in Sagamore Hills, Ohio. (See Author's Note at the front of this book for an explanation of how four different jurisdiction names refer to the one piece of land that was our homestead.) It is but one of a half-dozen small valleys that thread their tedious ways to the meandering waters of the Cuyahoga River on the valley floor, perhaps a half-mile in a straight line from the falls. It begins as little more than a low-lying natural ditch, once fed by a small spring long-since dried out, that drains a few gentle slopes in that nook of northern Summit County. For those few hundred feet it was known as "Jensik's Creek," so named by us for a neighbor who lived on its opposite side behind us but who, I don't suppose, owned any bigger piece of it than did the Bardow, Beers, or

Walls families further east. At about the place where our oddly shaped property came to a triangular point, Jensik's Creek runs over that small waterfall twenty or thirty feet high, the most arresting feature of the area's physical landscape and the one that deserved better than our blandly generic name for it. Impressive as it was to our young eyes, our little gulley was insignificant compared to the neighboring ravine a quarter of a mile to the south cut by historic Brandywine Creek and boasting magnificent Brandywine Falls. But Brandywine belonged to people like David Hudson, who, two centuries ago, paddled his way to a spot of wilderness that still bears his name; or, more recently, to the U.S. National Park Service. The falls—our falls—lower casing and all, belonged to us. But it was a child's sense of ownership.

"Belong" is a funny word, especially as it relates to land. In this country our use of it has always bordered on the laughable, if not the absurd. We shuffle a few pieces of paper among three or four people in some distant bank office then proclaim to the world that we "own" a piece of land a half-acre across, several thousand miles deep, and four billion years old. I once heard a descendant of an American Indian tribe try to explain how foreign such a concept was to the indigenous folks of North America: "You white people always talk about how easy it was to cheat an Indian out of his land," she said with a smile. "But to that Indian, selling a piece of land made no more sense than selling a piece of the sky. It never occurred to him that it was his to sell. He figured he had cheated you out of those beads and trinkets." And it would never have occurred to that same Indian that "selling" the land would make it any less accessible to him, just as selling a piece of the sky surely could not mean that he could no longer stand under it. Subjecting Creation's infinity of time and space to the stingy commercial rules of tiny, short-lived humans defies common sense. Doesn't it?

Our ownership of the falls and that ravine was an Indian ownership. If, as rarely happened, a grouchy neighbor shouted us away from it we would acknowledge his adult authority and leave it quickly enough. But it never occurred to us not to come back. It was there for either everybody or "one body," and the latter alternative seemed an absurd waste. Had my parents persisted in their early command to stay away from it, we would have had to disobey again, sooner or later.

Every kid has some piece of the world, no matter how silly or small, that is his or her special place. It is usually an obscure bit of nothingness that the adult world ignores, probably because it can't make any money off of it: a rotted stump, a basement corner, an abandoned car. The only two prerequisites for such a special place are that no one else knows of it (or at least has no interest in it) and that it unfailingly provides a dusting of magical charm. Our special place sprawled out over 2,000 acres and probably had not changed appreciably in the 20,000 years since pieces of the Wisconsin Glacier died a warm, watery death prostrated among its ancient hummocks. It is now part of Ohio's only national park, but at the time most of the adult world just passed by it. "The woods," stretching south and west from the falls, offered us a child's infinity of time, space, and salamanders. No matter how hard we pounded on her earthen shoulders, or screamed into her earlike caves, or pulled her grapevine hair that tumbled carelessly from her treetops, she only smiled sleepily and went about her ancient business.

A child's special place also offers escape. Under the steps or behind the curtain, he can ride the winds of imagination to jungles untouched by the human eye. But our special place actually took us halfway there in physical reality. We didn't have to wholly rely on the delicate cocoon of imagination that is so easily torn by housekeeping demands or bad moods. Like Twain's Mississippi River, forever providing Huck Finn with an avenue of escape and freedom, our canyon offered us a retreat from the small furies of our lives. It was magnificently consistent: always the same sights, sounds, and smells, and always there. It was also our broad-shouldered buffer against our greatest social fear, encroachment by the tide of city dwellers exiting Cleveland. Brushing aside any hypocrisy that should have arisen from our own urban flight, we kids looked with disdain and discomfort upon the rapid residential development of our rural morsel that sat exactly halfway between Cleveland to the north and Akron to the south. Lloyd, in particular, was constantly grumbling about the growing number of housing developments splaying themselves out on fields that had hosted crops for 150 years, and malarial jungles for thousands of years before that. But we were sure that they could never take the falls. Who could build a housing development on top of that? Even when I now view the adjacent ski slopes of Brandywine or Boston I cannot bring myself to picture a bulldozer touching our hallowed crease in the earth.

Once, as a middle-aged adult, my eight-year-old heart drove my middle-aged body down the length of that ravine to its insignificant spillage into the Cuyahoga at the bottom of Red Lock Hill. I was slightly shocked to realize that, as a kid, I had never made the trip all the way down the valley. I guessed that this was because of the curiosity of youth that drew us to every tiny treasure along the way. The woods, falls, and ravine, not the Cuyahoga River, were our destinations in those days. The place was an end in itself. We had no reason to seek the creek's terminus. It was mid-January, and I found our special place in splendid form. A dusting of snow enhanced the old magic, draping my childhood images with silky, white cobwebs, as if untouched since I had left them a third of a century before. The creek, broken by millions of rocks and weaving its torturous way downward with the patience born of endless time, was about eighty percent frozen over. I played a continuous game with the thin panes of ice, trying to jump off of the groaning, cracking surfaces just before they opened under me, enticed by the dicey prospect that the water's depth beneath the ice could be either two inches or two feet. As always, the creek didn't mind. It would have ample time to heal the cracks or, upon a different whim, melt them away altogether. At intervals the stream broke into open water, each spot adding a musical backdrop to the splashing chorus.

In spite of nature's sometimes-sloppy housekeeping, I was amazed at how clearly this old mansion told its story, a story to which I had been oblivious as a kid. The Berea Sandstone outcroppings stacked outward over the creek below, with the longest thrusts of the plates at the top and the shorter ones cutting back beneath them, in opposing image to the shape of the ravine itself. The softer beds of shales, however, followed the slopes of the ravine walls up to the base of the sandstone, leaving the impression that a giant arrow had dented the walls of the valley. Both of these rock formations were obediently following geological house rules, with the tremendously resistant sandstone almost impervious to the slashing water, while the shales below eroded quickly. Everything neatly reflected the house's multimillion-year-old geological blueprint. Not so with the furniture. Over-hanging, snapping sandstones and sliding shales mixed with fallen trees to create a veritable junkyard at the creek bottom. But there was no sense of embarrassment about all of this. With little more than a shrug the valley continued on its way, feeling no need to tell the perplexed observer that housekeeping was not its main job.

The trek was surprisingly short, taking less than an hour to reach the Cuyahoga below, near the National Park headquarters for what was then the Cuyahoga Valley National Recreation Area. I was somewhat disappointed to see that the creek did not make a direct run for the river once it hit the Cuyahoga plain, but rather was routed through a culvert under Highland Road and thence to its final destination just beyond another roadway leading to the old Jaite paper mill. I winced when I saw the final ignominy, a concrete ditch that carried these sacred waters of Jensik's Creek and the falls to their tiny mouth, as if the old lady hadn't been quite able to make it on her own and needed the strong arm of the rest home attendant to complete her trip to the place where she could expel her waste waters.

But then, just where the stream made its ancient contribution to the Cuyahoga River, I was rewarded by a wonderful sight—unremarkable to any other eye, but beautiful to mine. About six feet before the concrete ditch ended at the river's bank there began an escalating build-up of mud-silt that grew to several inches deep at the point where the creek and river waters mingle. The concrete was totally obliterated, defeated, for those last few feet as the stream matter-of-factly reenacted a modest exercise it had been performing for thousands, probably millions, of years. There were bits of our back yard in that silt, as well as those of the Beerses, Jensik's, and Wallses, and the creek was depositing them here at this train station on the Cuyahoga as faithfully as it had done since a time before there were Knowleses and Beerses and Jensiks and Wallses. I frowned at the audacity of the boringly symmetrical concrete ditch, then smiled as I looked at the primeval ooze that was calmly remaking the creek bed, even if only a few inches at a time. Our special place was still winning the war against human ownership, human encroachment.

The land still belonged to no one. And everyone.

∼∼∼

About sixty feet before the ravine wall begins its northern descent to the stream bottom, there is, or was, an open field. It too slopes downward in quiet genuflection to the water gods below, but the grade is so gentle that Uncle Frank was never confused as to its intended meaning. My dad's brother, with his wife, Doris, and our three sibling-like cousins, Janet, Dave, and Jack, owned the larger chunk of our almost-communal piece of the five-acre homestead. Where some men see a cornfield or a pasture, Uncle Frank saw a ball field, a 1950s field of dreams which, while never hosting the continent's best ballplayers, managed to stage hundreds of the best hours of our childhoods. It was another gift of the land.

Looking at the open patch of ground today you wonder how teenage boys and grown men ever made good games upon it. A planted tree here and a garden there, fringed by nature's incessant encroachments, have conspired to give a mythical quality to my remembrance of our competitive sport upon it. Only the rough diamond shape still remains. Yet, the games did go on, hundreds of them, in nine of the twelve months of the year, baseball and football. And in the later years of our childhood, they were the lively, competitive games of reasonably good athletes. Of course, we had to make a few adaptations. The first was for the condition of the ground. The slope of the land, coupled with Uncle Frank's miserly refusal to hire a full-time groundskeeper, meant that the rains were forever cutting tiny trenches in the infield. I don't remember one grounder that took all true hops on that infield. The ground balls came at you much the same as the creek below, ricocheting off of oddly placed obstacles in pursuit of open territory, which they often found between my tennis shoes. "Don't turn your head!" The remark was usually made by a pitcher, outfielder, or someone else whose nose and teeth were not in immediate jeopardy. When it came time to try out for the high school team most of the Knowles boys trotted to the outfield.

Then, there was the business of boundaries. Because the ball field was surrounded on all sides by woods, and because our supply of softballs was as small as the woods was large, we had to search for every foul ball. These were less than pleasant interruptions for us, though much appreciated by the mosquitoes, flies, and briars. Naturally, the farther the foul ball was pulled into the woods, the lengthier our trek and the longer our search. Usually the batter's first offense was met with stoical silence as we hoped that the third baseman or leftfielder would find it without having to call in the rest of the search team. A repeat performance drew a good bit of loud grumbling, and a third foul could make things downright ugly. I recall that Lloyd went through a batting phase (we all did, I suppose) where he was "bailing out" on each pitch, dropping his left foot far behind the third base line and, as a result, pulling inside pitches deep into foul territory. There were times on hot, sultry summer evenings after he had lofted a fourth or fifth foul deep into the woods when Lloyd's life hung in the balance. Some of his handiwork lies rotting on the soft floor of the woods today, perhaps yet hopeful—if softballs think about such things—of being found and rejoining the game.

Of course, the woods also extended around the edges of the ball field's "fair" territory, and it was this circumstance that created something of a "rite of passage" among us maturing boys in the neighborhood. Being able to reach the outfield woods with a fly ball meant manhood, pure and simple. I longed for the day when I, like my older cousin, Dave, could drive a softball completely out of this ball field which, to my little boy's eyes, had always seemed so big—to send it on a transcendent flight from our world of civilization back to the world of nature. My envy turned to absolute aching when Dave would casually step over to the other side of the plate and repeat the trick batting left-handed, especially since the right field woods line was the deepest of the three outfields. (Centerfield, oddly enough, was shortest, thanks to a tenacious tulip tree.)

But while parking the ball into the outfield woods made you a man, it also made you an out. The short fields, aided by the downward slope away from home plate, meant that any ground rule that rewarded such a drive would have turned our games into a chaos of lost balls and runaway scoring. Making the woods an out helped us produce both good games and good low line-drive hitters. I don't remember the first time I actually reached the woods on the fly. I suppose Dad may have sighed as he watched the ball carry into the first bush beyond the mowed grass, perhaps wistfully remembering the little boy who, at three, wasn't strong enough to turn the hands of the prize clock on the local King Jack TV Show. But his nostalgia didn't deter his "out" call. The rules of the land had to be honored, whether they pertained to creek beds or softball games. For my part, I learned something about the bittersweetness of accomplishing goals consisting more of form than substance. Within a few weeks of that landmark hit I was quietly grumbling every time I got the ball up too high and into the woods. It was just a noisy out, a glorified failure, with no more value than a pop-up to the catcher. Here was one rite of passage that, like shaving, promised much more than it delivered.

Our odd "rules of the land," if something less than sacred laws, were something more than capricious whims. There was a sense that we owed the land that much, that it was the least we could do. Perhaps it was an ineffective gesture—we were painfully aware how little we could do to properly honor or even protect this mass of earth that cradled us for so many years—but it wasn't phony. And there was a sense of ultimate seriousness about it.

Dad once gave a clumsy testimony to this unspoken honoring during one of our countless holiday ball games when the field was choked with visiting church friends and relatives. One of the visitors was a tall, sinewy guy named Dick, in the prime of his twenties who, if he never played competitive sports, certainly could have. Dad was playing third base when Dick headed there after a hit. On his way to the base Dick lowered his shoulder slightly in Dad's direction before heading for home plate. There was no contact between them, just that lowered shoulder. To our stunned horror, Dad made a beeline after Dick and took a flying leap onto his back about the time they both reached home. Dick, to his lasting credit, made no attempt to respond to this most uncharacteristic outburst from Dad, but just stood there, blinking nervously, waiting for someone to tell him what he had done wrong. There ensued about three seconds of shocked silence during which Dad, like the proverbial dog who finally caught the car, couldn't figure out what to do with his catch. Clearly, his leap had taken him far beyond the written script. Then he blurted, "Don't do that again, Dick. We don't play that way out here."

To most of our visitors that day Dad's behavior may have seemed strange, but to the seven or eight of us who had invested some of the best competitive moments of our lives on that field, who had for years been bringing to it libations of skinned knees and bruised hips but never a serious injury, Dad's response made sense, in substance if not in style. There was a special way to play ball *out here*, and it was necessarily a moral notch higher than the way sports were played at Garfield Park on a Sunday afternoon or Bedford High School on a Friday night. In those environments a lowered shoulder may have been common, even expected, but not on our ball field. Our ball field deserved something better because, well, because it had been so good to us. It had given us a theater for wholesome fun, a place where fathers could teach their sons sportsmanship, a rose-hued strongbox where memories could be stored forever. It was a giver of good gifts that asked nothing in return. And because it asked nothing we felt we should give it something. I guess our sense of respect for it was that something.

Along with our ground rules, we brought to our ball field a weird assortment of personal idiosyncrasies. As little boys some of these naturally took the forms of sports fantasies. Jack was Cleveland Browns fullback Jimmy Brown, I was quarterback Milt Plum, and Lloyd was (at least for a while) split end Preston Carpenter. We would crayon clumsy numbers, 32, 16, 44, onto old sweatshirts to help the magic along a bit, then slip into whatever Santa had most recently added to our growing stash of football paraphernalia. Jack was the first among us to get a helmet with a face bar on it, this apparently in response to the previous Christmas at which time Lloyd and I got complete uniforms (but the helmets had no face bars). Because Jack fell a bit short of Jimmy Brown in size, speed, and color, he tended to specialize in the one Jimmy Brown imitation he could manage: never giving up on a play. At that time, before we were big enough to dictate rigid adherence to touch-tackle, the men would stop us by simply grabbing us or, just as frequently, picking us up in mid-stride. Jack's stock in trade was to keep his legs going, bicycle style, while one of the men was holding him a foot off of the ground. His little legs did more work in the air than they ever did on the ground. For my part, I specialized in throwing safe screen passes. Like Milt, I never cared much for heavy pass pressure or high risk throws. Lloyd's Preston Carpenter consisted of an endless series of five-yard "down-and-outs." I suppose, like all kids, we thought we were getting close to the real thing.

But the fantasies weren't limited to us kids. The human being is one of the few animals that carries a real sense of play into adult life. Dog owners and squirrel observers might disagree with that, but the former would have to limit their argument to wild dogs, not the lovable household critters that simply mirror the actions of their owners, while the latter would need to offer something more than the frisky romps driven by mating behavior. Besides, I'm not talking about spontaneous bursts of energy, but rather the premeditated, protracted play we speak of as "games"—the kind of thing that invades your mind during the last hour or two of an otherwise busy workday, or lingers on in mental pictures long after muscles have ceased to ache and bruises have metamorphosed through their kaleidoscope of colors. It is not only the frolicking of children but also of grown men, men whose jobs at the time made no provision for disabling injuries suffered on the field of play.

Dad and Uncle Frank were two of the human animals who did not lose their sense of play. On paper, they were unlikely candidates to be our playmates. Neither could draw upon the lengthy or well-orchestrated game experiences showered upon our society's current crop of kids, and I never heard either of them mention an instance when their father

> The ball field, backyard. Thousands of hours of pure pleasure and, sometimes, painful maturation, were to be experienced here over nearly twenty years. Just a good-sized opening in the woods, Uncle Frank saw the possibilities immediately. *Photo from author's collection.*

had taken time to play with them. Grandpa Knowles had been far too seriously burdened with a depression and a dead wife to play with his sons. Instead, the boys made do, as children will, with what circumstances provided, carrying on games of Cleveland street ball until the predictable neighbor's call to the police sent them scurrying through a well-marked hole in a nearby fence. Nor did Dad or Uncle Frank play sports in high school. No time for that luxury; Dad was completely on his own from age fifteen on, working, paying for a car, and completing high school.

Yet, there was a night, still pristinely clear in the memories of several older Knowles men, when we answered the call of a new moon, and a late night snow, and the ball field's standing invitation to us. I don't remember anything about teams or scores, or about jobs or school the following morning, or about the inconvenience of late night baths and snow tracked onto the floors. What I do remember is an eight-inch line in the snow, a reminder of Dad's long nose and the last second of his futile effort to stay on his feet before plunging into the foot-deep snow. And I remember the unrestrained laughter and the magical brightness of everything hours after the winter sun had set. But mostly I remember the unique camaraderie of play.

While we boys were working on our pro football fantasies, the fathers were entertaining a few of their own. Uncle Frank's was the "basketball pass," which, as the name suggests, was a push pass from his chest. Unfortunately, we were playing football, not basketball, and Uncle Frank never quite got the idea that footballs weren't meant to be thrown that way. My guess is that he kept it handy because he never responded very well to a furious pass rush, even in the relatively safe arena of touch football, and so wanted a means of quickly dispatching the hot potato that was bringing him so much negative attention. The problem was that these panicky thrusts sometimes came before his wide receivers had a chance to turn around. We were forever being plunked on the back of the head by the ball, after which he would unfailingly yell, "C'mon, wake up! Basketball pass!" Physical safety was our only reason for paying attention to these glorified laterals since a reception only netted us about a yard and a half. But I think Uncle Frank secretly entertained the notion that his invention could find some good use in the NFL given a proper chance.

Dad's contribution to our inventory of plays was the play-action pass. While it is true that this play has become a standard of football stock, Dad's dogged insistence on trying it on our ball field won him little praise from us. The problem, you see, is that we were playing touch-tackle, two hands anywhere, usually three or four men on a side. These rules and numbers dictated a total passing game, with our only runs coming on quarterback scrambles (and even these we sometimes outlawed when the numbers were too few). Except when we were very little boys and the dads were trying to simulate some actual game conditions for our amusement, we never, *ever*, used a handoff into the line. With two defenders standing at the line waiting for the ball carrier, one on either side of the center and needing only to reach out their hands to kill the play, the line buck made about as much sense as giving your opponent the first two moves in a game of tic-tac-toe. So, faking it was not likely to fool anyone over six years old.

Back in the huddle, Dad's voice became animated as he shared with us his great vision for the play. "OK, now listen, this is gonna work like a charm! Lloyd, you start by me at halfback, see. When I fake the ball to you, you double-up and go barreling into the line. They'll think you got the ball, see."

Lloyd and I would exchange quick, pained glances before rolling our eyes heavenward. There was no use arguing with him; we'd just have to go through with the wasted play. When Jack or Dave, or whoever happened to be the lucky defenders saw the tell-tale line-up on the other side of the ball they smiled knowingly and mentally clicked off the down. I never remember what the rest of the play was supposed to be, but it was always played out in a kind of slow motion, with the defenders waiting patiently for the thing to fizzle. Dad's post-play assessment was also always the same, and delivered in a kind of hurt tone that suggested that our lack of enthusiasm for the grand idea had cost us a sensational opportunity: "You didn't give him a good fake. Double-over when you go into the line, like you got the ball." Which meant, unfortunately, that we would have to run it again in another week or two. Somewhere in the recesses of his residual, boyhood imagination, in the same part where Jack's air-beating legs convinced him that he was Jimmy Brown, Dad's mind's eye was playing for him a movie of this play working to perfection. It was the only place it ever worked.

Somewhere along the line, during one of a hundred unremembered and otherwise unremarkable football games, the ball field allowed each of us kids to celebrate that moment when sons physically eclipse their fathers, when nature's cycle demands that the baton be passed. If only for a moment, we stood at the equinox, each of us heading in opposite directions for the solstices of our manhood, winter for Dad and Uncle Frank, summer for us boys. If the land had given us nothing more than that, a positive and playful environment for realizing what too many fathers and sons realize only when a habitually delivered blow is suddenly returned with greater power, or when someone has already stormed out of a home never to return, it would have been a gift adequate for a lifetime.

The passing of the ships of prowess could only have lasted briefly. I don't remember much about those "even" times. It seemed like all of our games were complicated by allowances of some kind; fathers making allowances for little boys, teenage sons making allowances for older men. We learned the art of assessing the available manpower and quickly calling out names that would make for fair teams. We all knew, at any given moment, the status of our various warriors: who had lost something more than a step, who could no longer be expected to guard whom. Some defensive coverage areas grew, others shrunk. We had long since dropped the defective practice of choosing up sides; the class was chemistry, not Phys. Ed. The curious process saw many firsts—not so much notches on our young guns as they were ascending height marks on an invisible doorway: the first time Dad didn't catch me from behind; the first time I realized Uncle Frank could no longer block me; the first time that team valor became the better part of discretion when I trotted over to Dad's cornerback spot and (not without appropriate embarrassment) suggested that maybe he should let me guard Lloyd while he looked after the fat neighborhood kid playing center.

I think of those times now because I well remember the time my own son, Mark, and I paused briefly on that same plateau. He was thirteen, larger than I was at his age and within months of shooting by me. I was no longer making allowances for him. For this one, sweet moment in our father-son lives we enjoyed the full force of competition. On the basketball court, (I was never silly enough to go near him on a soccer field.) I had slight advantages in quickness, muscle coordination, strength, and height. He had more stamina, better dribbling and passing skills, and more "fire in his belly." I was vaguely aware that soon I would slow to a trot and watch with a wistful smile as he pulled out in front of me for good. When the moment came I was able to feel, as Dad and Uncle Frank must have felt, a sense of satisfaction at having run a good part of the race with my son.

It is sometimes said that sounds never really die away, that they, like the black body radiation of the universe's Big Bang, are always floating around somewhere albeit in steadily weakening sound waves. I sometimes wonder if the old sounds of our ball field days circle back around the place from time to time. That may not be as ridiculous as it sounds. After all, where else would they go? The not-so-new-anymore owners of the land couldn't possibly appreciate them, and the falls and woods never cared much for football or baseball. Perhaps the echoes linger on warm summer evenings or cool Saturday afternoons in October, longing for the sight of our single-file parade of ball gloves or tossle caps coming down the path, wishing for new sounds to be sent out into the air behind them, just as aging fathers wish for sons to make new waves in the slowing wash of their own wakes.

∼∼∼

Thales would have enjoyed the riotous marriage between water and the land in the Western Reserve, Ohio's twelve northeastern-most counties. The ancient Greek monist, a close observer of nature, noted that water alone among the "root elements" (water, fire, air, earth) seemed to hold power over all of the others, easily evaporating into air then condensing back again, magically building up earthen silt beds at the mouths of rivers, dousing the hot flames of fire. His conclusion that water is the cause of all things almost makes for believable philosophy in our part of the country. Certainly you cannot begin to understand the land until you understand its eons-old affair with the water. Hundreds of feet below the surface the two lay in strange embraces as rivers never seen by man twist through the dolomites, shales, and clays vainly searching for oceans that disappeared hundreds of millions of years ago. Vast piles of sandstone protect lenses of cool, glacial residue waiting to surge up into some distant kitchen sink at the prick of a well-digger's drill.

The first task of those who live on top of the land is to meekly inquire about its aquiferous resources. City water, piped from the planet's finest fresh water supply, was still 20 miles and two decades away from Northfield in 1953. Digging a well was as much a part of a new home as was laying a footer, and the kind of luck you had with that venture determined a great deal of your future comfort or frustration. Uncle Frank and Aunt Doris had a well that produced what we simply called "iron water," a liquid with such a strong mineral content that the mere thought of it was enough to drive away a hot, August thirst. All of Aunt Doris's cups and pans were tainted with the reddish film left by the stuff. I used to wonder if Cousins Jack and Dave would grow up to possess supernatural physical abilities for having imbibed so much iron water over the years. I suppose it wasn't iron, but rather some other acidic substance that tends to foul the eastern Ohio shales, or perhaps even a petroleum taint of some sort. There are stories of exposed shale deposits in the Western Reserve that burned for days after catching fire.

Dad had better luck with his well, probably tapping into one of the Berea Sandstone lenses that were so uniquely plentiful in Northfield Township. We got water at the very good rate of eleven gallons a minute, and the water was good. (I was always secretly proud of our luck with the well, assuming, as kids will sometimes do, that it all had something to do with moral character and God's favor. The thought was particularly welcome when I was mad at Jack.) Just to be on the safe side, a place Dad sought with regularity, he put the family through the ordeal of "softening the water" once a month or so, another rite of our water worship, this one involving salt pellets, a tank, and washcloths draped over faucets and the toilet handle with the standing admonition, "Don't use the water."

There was another reason we had luck with our well. Before drilling, the plumbing contractor used a divining rod to locate the best water source. Sure, I know; just so much hocus pocus, you say, an old wives' tale. Even the word "divining" draws up images of pagans predicting future events based on their readings of animal entrails. And the divining rod was nothing more than the "Y" branching of a willow sapling, the idea being to walk across the ground holding the "Y" arms until the stick's shaft pulled toward the earth to announce its discovery. The problem with a reasoned skepticism here is that the willow-witching trick worked. To this day, Dad matter-of-factly acknowledges that he felt the pull of the stick in his hand right over the spot where we drilled, probably not five feet from Uncle Frank's property line.

Thales would have understood. But I am not sure he really understood the relationship between water and the land. If those two ancient elements give evidence of the "oneness" the early Greek monists were searching for, they also contribute much evidence to the contrary. At times, they seem to be at war with each other. At other times, they stage a reasonably good imitation of *Dr. Jekyll and Mr. Hyde*. Those wells, for example, are anything but neat, aquatic cells packaged at regular intervals in the well-ordered crust of the earth. Rather, they are wildly strewn about the subterranean battlefields, oddly distorted shapes lying where they fell when the geologic wars passed over the earth eons ago. Poring over the soil maps, I see that Dad drilled his well right on the edge of the good, sandstone bedrock. One or two hundred feet to the southwest the land turns to lucastrine silt and clay as it falls toward the Cuyahoga River, substances with little talent for preserving water. Perhaps Uncle Frank's well strayed too far in that direction. But the water's capriciousness does not end there. Down on the valley floor of the Cuyahoga, the old Jaite Paper Company drilled for water in 1917 in order to meet their industrial demands. They finally struck water *450 feet* below the surface, and even at this depth, probably eight times the depth of our well on top of the higher ground, they did not strike bedrock. Their two wells, twenty feet apart, produced one million gallons of water a day until the late 1940s, at which time additional pumping power was added to increase the water production to two million gallons a day. Underneath that river bed was a deep, rich level of pre-glacial sands and gravels, tens of millions of years in the making, that hid tremendous water reserves.[1] But who would have guessed it? Perhaps the divining rods. Maybe that is one reason in addition to gravity why the trees of the ravines lean and eventually fall in the direction of the river.

If these two, land and water, are at war, the jury may still be out determining the winner. Only a handful of geological years ago, perhaps as few as 25,000, the land of my childhood was submerged, that time under the waters of Lake Cuyahoga that formed when confused rivers bumped up against the walls of the sweating Wisconsin Glacier as it slowly retreated northward. Even after its withdrawal, the glacier left some of its forts in place, such as the magnificent Portage Lakes on the watershed at Akron, and the Great Lakes basin. The ravines and hummocks of our part of the state, the eastern Allegheny Plateau, put up a much tougher fight against the glaciers than did the more submissive flatlands in western Ohio. The ice invader became bogged down in the uneven terrain and had to content himself with the sabotage of great valleys. Even then the land laughed last as the billions of tons of earth being plowed before the glacial mass fanned out to become the moraines that gave the Western Reserve much of its bed of rich soil.

But the glacier, four of them, actually, was a newcomer in this war of water and land. A half-billion years earlier the earth had begun its cycle of vast inland seas dried by rises in the land surface followed again by deluges. Who can say whether northern Ohio's geologic history was spent mostly above or below the waves? Probably the latter, given what those massive layers of sandstone must have required from the ticking of the sedimentary clock. At any rate, even in modern times there is something of the sense that the water is only waiting, with the infinite patience of time not troubled by human limitations, to rise up and again reclaim the surface. Sometimes the hints are softly murmured, like the tiny shell fossils embedded in the rock I pulled out of the garden a few years ago. Other times the old water warrior is not so subtle. The sleepy Cuyahoga, at times a mere thread of wetness on the valley floor, can become an instant maker of an inland sea when the rains fall and she calls in her tributary chips. Newspaper accounts of the great flood of 1913 stated that the river was a mile wide in some spots. Even during normal rains the Cuyahoga has a way of swamping the areas around Independence, just to the north of Northfield. Our Sunday drives to church down Canal Road were a wonder for me at such times, tantalizing hints of what it must have been like when "darkness was upon the face of the deep; and the Spirit of God was moving over the face of the waters."

Walls's Pond hardly qualified as an inland sea, but as kids we might have argued the point. Probably no more than fifty feet rim-to-rim, the pond was nothing more than an arbitrary low spot close to the point where a natural spring once gurgled forth to became Jensik's Creek, then the falls, then the ravine creek, then the Cuyahoga River. I don't suppose it was more than three feet deep in any one place. It was, in fact, such an insignificant body of water that a house now sits on its old bed, compliments of what was probably a few, routine swipes from a bulldozer.

But for us, Walls's Pond was an aquatic wonder, especially in the springtime. In late March, when the ball field was still a sea of mud and the woods slippery with cool stacks of damp maple and elm leaves, we would troop down to the pond to watch the pulsating reawakening of life. The waterstriders' Olympic level routines on the smooth surface was usually worth a moment's time, but soon enough we were stepping gingerly into the cool waters in search of bigger game. The frog eggs, tiny black beads suspended in gelatinous sacks, told us we were on the right path. But these we did not disturb. The real fun came when we caught up with the polliwogs, those all-head-and-tail critters (libeled in dozens of sex education films because of their close resemblance to swimming sperm) that so little resembled their parental frogs. Because they were so quick and never swam in a straight line, catching them became intriguing work, and could turn the hours of a Saturday morning into what seemed like fifteen minutes.

Once we brought a bucket of polliwogs home for further investigation. I was awed and excited when I noticed one day that a couple of them had sprung tiny back legs. But something wasn't quite right. The little guys seemed to be languishing for some reason, despite our best efforts to provide for them. Mom and Dad smiled and shook their heads, wondering aloud for our consideration if frogs were meant to live in buckets. After one of the polliwogs died, we returned the remainder of the captives to the pond. I remember a similar cycle with a turtle we once caught. I began to suspect that he wasn't eating properly when he took a chunk out of my index finger. When we gave him his unconditional release into Walls's Pond he immediately stuck his head under the water and into the mud. So much for the presumed superiority of human hospitality.

Maybe everything needs to return to the water once in a while. Maybe, like that turtle, which we found on the ball field hundreds of yards from the nearest water, we are only straying on the long leash the water indulges us. Loren Eiseley, renowned naturalist and gifted writer, spent a good part of his life kicking around the Walls's Ponds of the world, wondering about the origins of life. He said that at some obscure moment in some primeval swamp we came ashore forever, leaving the water for the land. Some intelligent people dispute the likeliness of that scenario, but the ancient call of the water tells another story. In the Western Reserve, that call is everywhere.

~~~

A river flowed out of Eden, we are told, and thence divided to drain the four corners of the world. We, too, had an ancient river flowing out of our paradise in northern Ohio, though there is some debate as to its name and how it got there. Some say it was the Dover, other suggest a separate river more appropriately named the Akron. In either event, the river was strong and deep, slashing out the basic contours of the Cuyahoga Valley long before the glaciers came. To the south, the prehistoric Teays River ran a mighty course, as the pre-glacial rivers were wont to do, draining everything from North Carolina to northern Indiana, and carving out part of the Ohio River bed in the process. Much of the rugged beauty created by this rivering of Ohio is plainly visible today at places such as Cuyahoga Falls and Brandywine Falls, and its magnificence teases us with the ancient question: Was this the deliberate design of a creative hand, or rather the play of nature's capricious whim?

In C.S. Lewis's *The Magician's Nephew*, the author offers an interesting variation on the creation story. He portrays Aslan, the great lion who represents Christ in the land of Narnia, as creating the world through a song, or a series of songs, more exactly. Each song the lion sings becomes a carrier of creation, spinning off thousands—millions—of lesser creations in riotous celebration of its new found license. Here, God is not the careful craftsman, poring over the details of his puppets in a dimly lit workshop, but rather an explosion of physical and spiritual energy blasting his creative power to the farthest reaches of whatever was meant to be. In the echoes of creation's songs, trees sigh, flowers bloom, and life crawls. Henceforth, everything on earth would lean in the direction of that creative song dancing just ahead of the wind over there, even if it meant that rocks and stones would sometimes cry out in their pursuit of it.

Creation's songs passed through the Cuyahoga Valley, too. Only instead of sound it was a symphony of silence broken only by the wind and the waves. And the rhythm was so slow, so infinitely slow that twenty successive human generations could not have detected a single beat of the earth's heart had anyone been there to hear it. They would surely have been driven mad by the sameness of it all. But the music was there nonetheless, in the heat of the sun, the steady pouring of the rains and rivers, the stiff pushing of the wind. Imperceptibly, these natural elements were wearing away the monstrous igneous rocks of the earth's fiery creation, scouring the granite crystals of quartz, mica, and feldspar into sand, gravel, and clay. These new rock by-products were washed, rolled, and blown to the broad but shallow inland seas where they made neat and orderly deposits, sediments, on the ocean floors, with the heavier sands falling first (that is, nearest the shores), while the finer materials drifted further out before settling to the bottom as clays. When the lands heaved up, the sea bottoms became our bedrock of sandstone, shale, and limestone. Geologists tell us that most of northern Ohio's significant sedimentary rock formation occurred during the Devonian, Mississippian and Pennsylvanian periods of the Paleozoic Era 250 to 300 million years ago.

~~~

Many of my Christian sisters and brothers recoil when the discussion begins taking this turn. There is something about the vastness of those endless millennia that seems like godlessness. Two hundred million years devoted to the cementing of sands and the rolling of pebbles appears a sacrilege against God's purposefulness. We insist on a creation in which God spins the cloth only long enough to give him the necessary material to cut out his ragdolls. Let fossils and sandstones be hanged. If we must explain their existence, let's just say that God made them to look that way, as is, when he made his very modern Adam, navel and all.

The problem is that this sense of outraged sacrilege comes not in defense of God who is not, I think, very much threatened by the physical evidence of his creation, but rather in our defense of time. Or, more accurately, our distorted ideas about time. Far better theologians than I have concluded that there need not be a problem between Genesis and geology: that, indeed, the creation cycles outlined in the bible strongly hint at a very similar story that the earth was telling me as I walked the floor of the ravine on a winter day in January. But there is a bit of the Bishop Ussher in all of us that cries out for a world that is less than 10,000 years old, which progressed directly from Eden up the Euphrates to Babylon, thence to Assyria, Egypt, Israel, Greece, and Rome. We declare that time is short not because it is short for God—not very likely—but because it is short for us. Eighty years makes a dent in 7,000, but not even a vapor chamber trace in four billion.

It was Darwin who first fired the world's imagination to the possibility that, given enough time, there was a plausible scientific theory for at least the living things of the world. His contribution was not the idea of evolution—careful observers of the world's critters had long since come to suspect something along that line—but rather natural selection, one of the great scientific theories of the modern world. A million nineteenth century Christians, uneasy with the anti-intellectualism that so often accrues to religions, must have silently sighed and concluded, "Yes, that makes sense. He's probably right about that." Worse, they secretly wondered what else he might be right about. They had lived in fear of this "other explanation" out there, the one that would one day expose their faith as a fraud, and now it seemed about to happen.

Only it never did quite happen, at least in the neat, compartmentalized way it was envisioned. Alfred Wallace, credited with co-discovering natural selection with Darwin (the latter had timidly waited nearly three decades to publish his theory, doing so only when prodded by Wallace's new found insight that threatened to scoop his great idea) suggested that when it came to the development of humankind, the hammer of natural selection wasn't heavy enough to smash the last puzzle piece into place. There was nothing in natural selection convincing enough to explain humankind's explosive brain development in the last 100,000 years; no linkage of fossil evidence to indicate that the thing had happened as a natural part of the competitive/adaptive process. Worse, the seemingly primitive natives of earth's far corners, in whom nineteenth century anthropologists hoped to find the "missing link" verification of Darwin's theory, proved to be not so primitive. In fact, Wallace contended that their primitiveness was social only, and proceeded to shock and disgust his fellow Victorians with the conclusion that the brain development of these "primitives" was about as advanced as that of anyone else, including the Royal Family. This was more than a class blow to the British world at its zenith, for it implied that the rapid brain development of these peoples had occurred even when it wasn't necessary (i.e., during a time when the natives were making little use of their burgeoning mental capabilities), a refutation of the primary tenet of natural selection. (Even today, Neo-Darwinists are torturing logic in desperate attempts to show how, conceivably, some aspect of emerging human behavior—perhaps the need for coordinated hunting practices—could environmentally explain this unprecedented advancement in the human brain. Funny, how unscientific we can become in defense of science.)

Wallace broke Darwin's heart when he began searching for divine authorship of this mystery play.[2] Yet, the ensuing century and a half has neither allowed Darwin to rest peacefully in his grave—his current defenders have seen the need to retreat behind that hyphenated "Neo" before the Darwinist label—nor to abandon the idea that creation is inextricably linked to a sense of divine purpose. Nevertheless, the theory is sound—at least as sound as our theories can get. Any theory that has successfully survived the past century and a half of unparalleled scientific scrutiny has at least one foot grounded in truth. The problem is not in the theory; it's in the presupposition, the unproven assumption that purposelessness is at the heart of creation. In other words, we are not allowed to have faith in the proofs of natural selection unless we first have faith in the assumption of philosophical naturalism and its implicit atheism. It's an odd rule, and one rightfully rejected by many scientists, Christian and otherwise. So, too, by me.

Years ago, during one of our countless picnics at the Knowles homestead, I had the good fortune to get Uncle Herb talking about his idea of heaven. He had been a minister in the Christian Church for many decades, but his comments about heaven that day would have probably earned him a rebuke in many bible colleges in the land. "You know," he said, "I think that maybe heaven is going to be right here on earth. We keep on making things better in this world. I think God will give us time to bring in the Kingdom here. Physically here."

Somehow, the scenario sounded suddenly believable. This surprised me, for the words, in themselves, were ridiculous. This was the 1960s, after the "War to End All Wars," after the war that followed and dwarfed it, after the Depression that robbed an entire generation of its subsistence and its dignity, after the realities of the Atomic Age left us dangling on a thread. Heaven here? This goodly and Godly man was pushing his luck. Yet there *was* a ring of truth in the statement, and it had everything to do with the sense of direction implied in it. We are, Uncle Herb was saying, going somewhere. Somewhere good. Creation *is* directional.

That was apparently Wallace's reverential judgment as he pondered the unexplainably large cranial capacity in the skull sitting in front of him on his laboratory table. So, too, perhaps for the renowned anthropologist, Loren Eiseley. But Eiseley resisted the jump from secular to sacred. His emotionally scarred childhood and lengthy hobo forays aboard western freights taking him nowhere in particular apparently did little to convince him of a loving God. Nonetheless, he had picked up the same trail Wallace had stumbled upon, and the honesty of the scientist within him would not let him call off the hunt. Late in life, at a time when other superstars of science were speaking at museums named for them, snaring huge federal research grants, or awarding each other with prizes at scientific conferences, Eiseley was still kicking around ordinary ponds at the edge of urban sprawls, searching feverishly for the directional secret of life. His own relentless, driving quest mirrored that of the secret force he pursued. *The Immense Journey* is as much a reflection of his own search for that creative force as it is the documenting of Man's progress from pond creature to computer builder. As a scientist, he readily acknowledged this force as an "organization," but his description of it hints at something more than capricious chance.

> Yet this organization itself is not strictly the product of life, nor of selection. Like some dark and passing shadow within matter, it cups out the eyes' small windows or spaces the notes of a meadow lark's song in the interior of a mottled egg. That principle—I am beginning to suspect—was there before the living in the deeps of water.[3]

The essence of that suspicion is the story of the land of my youth. That is why I felt so magnetically attracted to her long before I knew the reason. The land is more than the footprint of God, for footprints only fill with water and outline themselves into stone. They only tell you that someone has been there. Rather, the land is God's very breath, warm, moist, alive. And it is still *moving*. Uncle Herb, the Christian minister, sensed that movement, and so, too, did Loren Eiseley, the secular anthropologist. And it is moving back toward its creator.

Small wonder that we who lived on such an unblemished piece of God's handiwork were transformed by the journey.

Notes

1. George Willard White, "The Ground Water Resources of Summit County Ohio" In Glacial Geology of Northeastern Ohio, 49. Columbus: Ohio Department of Natural Resources, 1982.

2. Loren Eiseley, The Immense Journey (New York: Vintage Books, 1957), 79-80.

3. Eiseley, Immense Journey, 26.

2
Valley Life Line
The Canal

Wilson's Mill, Canal Road. Also known as Alexander's Mill, for the 19th century owners, was using water power from the canal lock as late as 1969. Lock mills made use of the ten-foot fall at each lock by cutting a separate channel from the canal high side and dropping the flow over a water wheel that supplied the power for the grist milling. The low side water was then returned to the canal. The building is still being operated as a feed mill.

Photo by Ian Adams.

The story of a nation's life line • The ghost that won't go away • Faint imprint on a forest floor

Every place has its ghosts. So did our neck of the Cuyahoga. But real ghosts are not the silly creatures of stories, alternately lovable or malevolent, and always a bit frightening. Neither are they necessarily human in form. Real ghosts are, in the older sense of the word, the restless spirits of those things that the world has not properly buried. Something unavenged. Unfulfilled. Or unappreciated. The thing's final chapter was never written, leaving behind a tormented spirit drifting restlessly across the land, pursuing its final justice not with frights or cruel tricks but with gentle sighs and soft reminders. These usually go unnoticed by the world that forgot to finish their graves.

As a boy, I regularly rubbed elbows with the sad ghost of the Ohio & Erie Canal. Of course, I didn't recognize the watery specter for what is was, but odd hints of its presence touched me from time to time. Red Lock Hill, for example, was as much a part of our landscape as the ball field and the falls. Sweeping steeply down to the Cuyahoga Valley floor, the hill was a severe test for school bus drivers in January and bicyclists in August. One of my favorite primary school-age visions, endlessly redrawn on the blackboard of my imagination, was the way our bus driver made his turn around at the bottom of Red Lock Hill, backing the hulking, yellow monster into a slim driveway posted on either side by concrete pillars of some kind. To my seven-year-old mind it was an exquisitely neat trick.

The driveway could have been the one belonging to the Jaite Paper Company, hidden from sight but around which the tiny community down there had been built. Or it may have been the driveway that led to Bush Joe's place. Bush Joe was a kid so nicknamed because his mother used to send him to school protected from the winter elements by a babushka over his head. Joe seemed to sense some need to be embarrassed by our comments—more so than we perceived our need for sensing shame in mocking his family's poverty and probable immigrant status—but he wore the babushka anyway and usually with a wide, innocent smile. I thought the people down at the bottom of Red Lock Hill were a bit strange.

For thirty years it never dawned on me to wonder why the hill wore the name Red Lock. Nor did that question occur to my brother, parents, aunt, uncle, or cousins living on the top of it. We did not see the ghost in Bush Joe's homely, round face or smell it in the stench regularly produced by the paper mill. Ghosts aren't quite that obvious. But had we known what we were looking for we might have sniffed some slack water in the air. There was, however, one part of the ghost—or, perhaps I should say an exposed piece of the dead body—that remained readily available for our unused inspection. In the 1960s, a rare, extant section of the canal of a mile or two began just to the north of us and ran as far as one of the Cleveland steel mills that was only interested in the water the ditch conveniently brought to its doors. The highway that accompanied the waterway along this humbling trek was dutifully named Canal Road, a route we traveled every Sunday when we were still commuting to a city church in Cleveland. At the point where the canal and the road separate, the ribbon of water dances coyly around the back of a hill and out of sight. *I wonder where it goes. I wish I could get out and follow it*, I would think to myself, tantalized by the prospect of a storybook land where canals go on forever. Even many years later, as a presumably well-educated adult, I did not make the connection that the canal ran back to the bottom of Red Lock Hill (or, at least the path of it did), then right through the crumbling lock, itself.

Red Lock, the thirty-fourth of forty-four original locks between Akron and Cleveland, was so named because a long-forgotten lockkeeper painted part of it red. One hundred and seventy-five years ago everything about the canal was intensely personal. Mere numbers would not do for locks any more than they would do for a settlement (and, in fact, the two names were often synonymous); there had to be a name. Red's neighbor locks sported handles such as Johnnycake, Mud Catcher, Pancake, Lonesome, and Whiskey. Old Red had to settle for a color, but the canallers knew right where—and who—she was. At some point the steep hill behind the lock picked up its name.

The lock and the stretch of the Ohio & Erie Canal at the base of our hill danced along a hotbed of history. This land played a role in the nation's great Indian wars, treaty lines, and interior economic expansion. Northfield Township was chosen as one of the first work sites for the canal laborers, and when they began digging Red Lock they stumbled on a rich burial site that yielded an early French trade gun contemporary with what were probably the first white people in the valley. But all of that was buried under a shaggy covering of trees, weeds, and moss by the time we kids made our childhood sojourns through the valley in the 1950s. The ruins of the lock escaped our usually intense curiosity—for many years I assumed it was a decayed bridge of some kind—while the old canal bed was nothing more than a faint, wet-leafed depression wandering through a woods no one traveled.

I think that is where the canal ghost was born—there among the choking weeds twenty feet north of Highland Road. What I assumed to be the concrete abutments of a long crumbled bridge were actually the magnificently sculpted walls of huge sandstone blocks which, unmolested by humanity's fickle nature, might have preserved those locks forever. For half a century, the wealth of a nation flowed through those walls, coursing a rich supply of economic blood to Akron, Cleveland, New York, and New Orleans. But for two generations before our arrival they had not received even a passing glance from the people in the speeding cars. Those cars, along with the busy Baltimore and Ohio Railroad trains across the river, were the respective grandchildren and children of this mother of transportation in Ohio. But in their selfish haste they had neglected to properly lay her to rest.

~~~

In 1820, there was another boy who grew up on or near our property in Northfield, and he had everything to do with the forgotten story of the canal. I can't prove he lived there, but probability at least favors the prospect that someone very like him did. Northfield was well established by the second decade of the nineteenth century, and Moses Cleaveland's surveyors had marked it as one of the Western Reserve's four most valuable pieces of township property as early as 1797. The land was in use by 1820, and farmers in those days had good reason for wanting lots of sons. I can almost visualize my historical counterpart who predated me by 140 years or so. His name might have been Samuel.

Samuel and I had shared some common knowledge. We both knew how the ice hangs like stalactites from the falls in January, and how looking across the Cuyahoga Valley from the top of Red Lock Hill always leaves you with the odd sense that you are looking both down and up at the opposite wall of the valley. But the land was about all Samuel and I shared across a century and a half. His life, and that of his family, were brutally difficult in pre-canal northern Ohio. They had, like most of the others, come from civilized Connecticut, but here they were reduced to living like exiles. The problem, in two words, was cash and distance. What was the point of doing the back-breaking work of clearing this rich land, tree by tree and stump by stump, if there was no place to send your crops other than to your own stomach? William Ellis, author of the landmark work, *The Cuyahoga*, highlighted this economic Achilles' heel that was crippling the Western Reserve in its infancy:

> The trouble was you couldn't get a barrel of, say, Ohio flour to New York overland without paying a dollar a hundredweight drayage every hundred miles to push it across the Appalachian barrier. That put a barrel of Ohio flour on the New York dock for about eleven dollars—and there she'd sit.[1]

The only alternative to wagon transport was water. In my mind's eye I can see Samuel's father deciding to try the river route to market. One time, and one time only, he probably opted to float his produce south to New Orleans, the seemingly better option open to farmers in the southern part of the state. North, down the Cuyahoga to Lake Erie, made more geographic sense, but Pa was leery of the fever that was again raging in

Cleveland. Besides, there was little reliable shipping traffic on Lake Erie and no place to ship anything to once you got there. Unusually heavy late summer rains, which lifted the shallow southern Ohio rivers to a navigable level, and Pa's growing desperation for cash had convinced him that the southern route was a better risk.

It wasn't. After a difficult portage over the old one-mile Indian path at Akron, which could have been much worse had Pa not chosen a time when the rain-swelled rivers had shortened the normally torturous eight mile portage, and the difficult task of knocking together several rafts to carry the flour, whiskey and pork, they began their float down the Muskingum-Tuscarawas River chain. The higher waters that kept the rafts from stalling on the river bottoms also created currents and wisps of whitewater that bounced poorly secured barrels into the water, sprayed the remainder with a continuous mist that hurried spoilage, and routed the preoccupied men and boys into the hammering limbs of overhanging trees. A quarter of their produce and a good portion of their endurance were used up before they ever reached the Beautiful Ohio.[2]

The trip down the big rivers was more restful, but still not fast enough. At times, Samuel thought he could smell their cargo beginning to rot. Their arrival at New Orleans proved a fresh horror. Because hundreds of other farmers had the same idea as Pa, the docks were glutted with the produce of America's interior drained by the Mississippi and Ohio Rivers, driving the prices down to disastrous levels. Samuel's family had to hang around for several days until the market prices hit a mark that might keep the family just this side of bankruptcy. In the process, they squandered more of their precious pennies and lost many gallons of the whiskey to thieves during their first unsuspecting night in town. Their eventual sale was made in the notes from some unfamiliar bank, but Samuel's father assumed that it was all right since, hey, money was money. But in those wildcatting days, each bank's notes were separately discounted. Only a professional financier knew the daily value of each. A kindly passerby informed Samuel's family that the notes with which they had been paid were among the least reliable in New Orleans, and that they were, at that moment, losing face value literally by the hour.

With nothing to show for their year's crop and the agonizing journey, the men of the family began an even worse nightmare: the trip home. Because rivers have a habit of flowing in only one direction, and steamboat tickets require cash, the men and boys set out on the longest walk of their lives. They had a few dollars, thanks to the barge logs that they had been able to sell for lumber in New Orleans, but there was little else to help them deal with the turning weather and great distances. One of the older boys, who at fifteen was about ripe for the world anyway, decided it wasn't worth the effort and stayed in one of the Louisiana river towns where he saw a pretty girl and a slim chance for a job. The rest of them didn't get home until almost Christmas. Meanwhile Pa, extremely vulnerable to the lure of a quick buck, had been fleeced of his remaining pennies in Cincinnati. Because the winter was not too hard, they eked out an existence of sorts on the garden stuffs carefully preserved by the women of the household. The several gallons of whiskey still stashed in the pantry also came in handy, especially for Pa. He was never far from the jug after that trip.

With the economic collapse of 1819, the destitute farmers in the Western Reserve were firmly iced into their cashless desperation. Everything squeaked to a tight stop, including trade, immigration, and land sales. There was no prospect for hope, no chance that they would ever have money for windows and stoves and plow parts and basic tools. They were less well off than their few remaining Indian neighbors, faced the certainty of failure and—come the first really bad winter—starvation. Samuel and his family were out of options. They were doomed.

Until the canal came along.

Actually, two canals. Dewitt Clinton's more famous Erie Canal brought the Eastern Seaboard's voracious appetite as far as Buffalo and the eastern reaches of Lake Erie. The Ohio and Erie broke ground in 1825, a year after the New Yorkers showed a largely skeptical world that the thing could be done (although Clinton had been impatiently, if temporarily, dumped out of the Governor's office during the long, expensive years of the difficult construction). Ohio's completed section between Akron and Cleveland celebrated its maiden packet voyage when the *State of Ohio* hauled into Cleveland on July 4, 1827. But that boat contained only the political bigwigs and their inexhaustible supply of hot air and ego. That same day there came a second canal boat, less noticed by the celebrating crowds but of vastly greater importance. This one was a freighter carrying wheat—wheat that the day before had sold for twenty-five cents a bushel but that the next day would sell for seventy-five cents. James and Margot Jackson report that wheat merchants from Buffalo, who had purchased only 1,000 bushels the year before the canal opened, were buying 250,000 bushels within a year of the *State of Ohio*'s ceremonial voyage.[3]

Instantly, *instantly*, Samuel and his family had gone from languishing survivalists to prosperous farmers. The effect was stupendous, and hardly to be appreciated by a generation that agonizes over a ten percent rise in the price of gas. It was the kind of cataclysmic swing that could only happen in the all-or-nothing world of the volatile frontier. Corn and flour followed wheat's heady price rise, increasing fivefold in the five years after that memorable Fourth of July, and another tenfold during the two decades that followed. Immigrants poured into the Western Reserve, 70,000 in the 1820s and 100,000 more in the 1830s. Property values rose 360 percent during the eight years after the canal was opened in 1832 all the way to Portsmouth on the Ohio. Cleveland and Akron, respectively a diseased swamp and Indian trail crossing in the early years of the new century, exploded into major American cities from almost the moment they were umbilicalled to the thin ribbon of slack water.

Samuel and his family saw the face of prosperity at last.

~~~

One of the great teasing wonders of the canal is that, physically, it appears incapable of authoring such a powerful chapter in America's economic history. The thing looks like child's play, a thin ditch scratched in the ground that hardly seemed big enough for one boat, much less two on opposite courses. The aqueducts needed to carry the waterway over existing creeks and rivers, such as could still be seen at Tinker's Creek in Independence as late as 2007 (and since replicated by the Cuyahoga Valley National Park). They looked like glorified erector set constructions. You got the feeling that a solid kick in the right place would collapse the whole thing. In one of the few canal novels to be found in modern libraries, *Chingo Smith of the Erie Canal*, the young protagonist, who had often fantasized about the great New York construction about which he had heard so much, reflects the almost universal disappointment that accompanied a first glimpse of a canal under construction. After months of hope that Dewitt Clinton's great dream might mean a dramatic change in his miserable fortunes, Chingo stumbles upon a group of men standing idly around in a narrow ditch filled to ankle level with green, fetid water. The orphan timidly addresses one of the men when he realizes that a question was not likely to interrupt anything important.

"Mister, what's that hole in the ground?"

"That's the canal, laddie," the man replied.

"The Erie Canal?" Chingo couldn't believe it. He almost cried with disappointment.

"Sure-ly. The Grand Ee-rye-ee Canal. All there is of it, except ten or twelve more miles just like it."[4]

Even after completion, the canal looked pathetic when drained of water. Anything from a violent storm to an ambitious muskrat could do the trick, especially where unscrupulous contractors had filled the banks with fallen trees and other debris rather than the solid fill dirt required by their contract. At such times, the mighty eighty-ton freighters and sleek passenger packets, slumped down in muddy puddles on the canal bottom, looked like toy boats left in an empty bathtub.

It did not take nature any great length of time, especially by her standards, to heal the scratch on her surface. Beyond the ruins of Red Lock there was, by the late twentieth century, only the faintest hint of a directional line heading north toward Cleveland. An unsuspecting hiker in the woods would miss it altogether. As Dad and I walked that line during a gray day in 1990, I thought of the melons I used to grow in my garden up on top of the hill a generation before. In June, the vines of those melon plants were rich and green, thrusting and climbing with so much vigor

Boats stranded along the canal in Massillon. Massillon was known as "the Wheat City" for the huge volume of wheat transacted at and through the town's canal connection. But canal boats can't run in the mud, as demonstrated here by a string of boats idled by a breach in the canal bank. *Photo provided by the Summit County Historical Society of Akron Ohio.*

that I had to reroute them back into the garden every day. But when fall came, the vines had become thin pencil lines on the ground, unmoved for weeks, now pumping every ounce of their waning energy into their pulpy fruits. It was a ritualistic, seasonal suicide; the vines would die so that the fruits could live. That was the canal's story. A teeming waterway cut through a primeval jungle, four feet deep, twenty-six feet wide at the bottom, and forty feet at the surface, now lay almost invisible along the cluttered carpet of the floor of the woods, while the oblivious distant fruits of her labor, Cleveland, Akron, Massillon, and other cities, continued to ripen along her vaguely shadowed line.

~~~

The canal story is, like most American narratives, a strange mix of persistence, personalities, and politics. Despite the early recognition by men such as Washington and Jefferson that a canal through the Ohio country was the only thing wanting to link up the Atlantic Ocean and the Gulf of Mexico, the idea was not popular during the early years of Ohio's statehood. Internal improvements of such a size, usually requiring the involvement of a strong state or federal government, were risky political ventures in the nation that had, less than four decades earlier, cast off a tyrannical central government. Politics aside, the costs also seemed prohibitive. Ellis noted that talk of a $5 million canal in a cash-starved state with only $133,000 in the treasury pushed even visionaries to the limits of their vision. There were no New Dealers in the Ohio legislatures of the early 1800s. Ardent canal men tended to have short political careers before 1825.

Necessity, the ancient mother of invention, joined hands with steely-eyed men like Governor Ethan Allen Brown, Cleveland lawyer Alfred Kelley, and Akron's General Simon Perkins to finally carry the canal from a wild idea to a hard reality. Even after the legislature approved it, the bitterness of the fight continued, with only a change of fighting rings. Erie Canaller James Geddes's early topographical studies, which favored the Cleveland to Portsmouth route that coincided nicely with the land interests of men like Perkins and Kelley, drew the hostile, nearly fatal fire of the people in Chillicothe, Sandusky, and the other more established sections of the state. Ohio banks refused to finance the project, and canal agents sent to New York to feel out the potential bond market found that the bypassed cities had taken out sabotaging ads in the New York papers to announce that their citizens would pay no taxes to support the thing.[5] Their strident opposition would continue until 1832, when the opening of the full length of the canal to Portsmouth forever doomed these early urban success stories to second-class cityhood in Ohio.

The political chicanery filtered down to all levels. The people in the struggling village of Cleveland managed to pull the rug out from under their larger and more prominent urban neighbor, Ohio City, on the west side of the mouth of the Cuyahoga, by swinging the canal terminus to the east side even though that option was several thousand dollars more expensive. That tiny nudge of a couple of hundred feet along the river sent Cleveland to world prominence and Ohio City to dusty archives and faded memories, leaving behind only a rather friendly eastside-westside rivalry for which few can remember the original cause. Perkins, meanwhile, was hop-scotching around the high ground that would become Akron, exhibiting only superficial and capricious interest in land that lay along the right-of-way of the continent's destiny while secretly wondering how many taverns and hostelries might profit from passengers waiting the many hours for their boats to thread their ways through the twenty locks descending the north face of the summit at Akron.

But it was men and women of lesser fame and fortune who midwifed the waterway down the birth canal. These were the hard-muscled sons and daughters of Ireland, a people driven to the backbreaking and heartbreaking labor on the canals by their need to work off the sea voyages that had brought them here. Many of these had already worked their way from the Hudson River to Buffalo on Dewitt Clinton's canal. Now they slid west to new dreams in Ohio as naturally as Connecticut families had done a decade or two earlier. With shovels, wheelbarrows, axes, and an occasional plow, they attacked a jungle that would have given pause to a bulldozer operator, and did so under the worst possible conditions.

A few, distant reflections of these conditions were known to me as a boy wandering the remnants of those malarial woods. The series of streams meandering down to the Cuyahoga created an almost limitless number of standing pools in which mosquito populations exploded to life. These would descend upon us in thick clouds when we dared to enter their dark, woodsy retreats during the day. In the evenings, it was their turn to come calling, turning our volleyball games into a strange choreography of jumps, slaps, brushes, and jerks, whether or not the ball was in play. None of us ever stood still between serves. Their constant assaults were enough to bring me to points of irrational hatred, the kind of rage in which I would sometimes surprise myself by gently catching one so that I could pull off its wings then turn it loose to suffer a slower death. Yet the remote ancestors of these unwelcome guests were far more troublesome for the bare-chested Gaels and local farmers undertaking the continent's greatest feat of engineering and physical labor. Untroubled by any defenses, the mosquitoes endowed their stinging tortures with doses of "Canal Fever," or "fever-n-ague," the malarial sickness that wasted men in the prime of their youth and left a series of shabby cemeteries dotting the banks of the canal from Akron to Cleveland. One of these, in Brecksville, hosts the remains of more than one hundred canal workmen.[6] Any piece of skin missed by the mosquitoes was claimed by the stinging flies whose descendants still conduct search-and-destroy missions on the Summit Metro Parks bike & hike trail on warm August evenings.

The local farmers appeared to beat the Irish to the punch when canal ground was first broken, but there is evidence that the Irish contractors allowed this to happen by design. The latter knew from long experience on New York's Erie Canal that the local contracts were based on artificially low bids that promised more than could be delivered—"Jesus clauses," they were called—contract provisions that Jesus, himself, could not have fulfilled.[7] There would be plenty of time and work for the Irish once these early contracts broke under the weight of unreasonable expectations and wishful financing. Ironically, some of the farmers, who would reap the greatest benefits from the canal, were initially suspicious of the project. They reacted angrily to an ad valorem tax reform that, not unreasonably, assessed higher taxes on the lands that would benefit most from accessibility to the canal. Initially, the locals grumbled that it was bad enough that state law already required two days road work each year from all males over the age of twenty-one, or a $1 tax in lieu thereof. But farmers are ultimately people of good sense, and they soon came to see that the canal would give to them much more than it would take. When the early contracts were opened in July of 1825, hundreds of them eagerly threw themselves into the great project.

And yet, a decent living and profits are not worth the farmer's life. Typhoid, smallpox, and cholera competed with Canal Fever to break the men by the hundreds and kill them by the score. Disease and the faulty contracts that the Irish-gang contractors had anticipated soon steered the project away from local farmers and toward the veteran Irish laborers. But while the contracts improved, the healthiness of the canal environment did not (a charge that would stick to the canal for all of its life). The Irishmen were philosophical about those they buried along the way. Death, like the mosquitoes of the air and the hard clays in the valley's ground, was part of the inevitable price to be paid for this job, and the assurance of a decent burial was the oddly important but singular request the men made in the face of it. Along the way, these toughened men occasionally felt the cooling stroke of God's hand. A nameless woman from Brecksville, two short locks up the canal from Red Lock, alighted on their sick camps and ministered to their burning fevers, earning for herself the name, "Angel of the Canal."[8]

For these hardships and risks the men were paid thirty cents a day, raised to one dollar if they were willing to work in water where rot and fungus would be added to their miseries. They were provided meals and lodgings in flophouse environments, but they at least had a chance to save money toward their passage debts if they could avoid the taverns that littered "Whiskey Lane," now Parkview Drive in Brecksville. Their bonus pay was a gill of whiskey each day that the men hopefully believed would keep the fever and ague at bay. For many of them, it must have provided the only hour of the day when the canal work made any sense. But sometimes the thing went beyond any kind of sense. At one point, the state had to bring in convicts from the Ohio Pen when others could no longer work. The brutal late summer of 1826 saw the project temporarily stopped until the heat retreated with the devils of disease.

Not surprisingly, men would often cheat in this dirty business. Occasionally, a contractor would simply duck out with his payroll funds, leaving his workers in the lurch. Those who stayed looked for more subtle savings. A favorite trick was to simply bury the bodies of the huge trees felled along the right of way in the canal banks, thus saving the precious energy and time needed to haul them two rods' distance to the woods line or burn them as called for in the contracts. The trick was not unknown to Alfred Kelley, the strong-willed canal commissioner who may have been the only man in northern Ohio capable of ram-rodding the difficult task to its completion on time. Kelley relentlessly walked the towpaths, prodding the ground with an iron rod in search of the hidden timber that would eventually rot and collapse the banks. Many people hated him for his rigid contract enforcement. Others wondered what kind of fool would risk (and often lose) his health in all kinds of weather for three dollars a day, his eyes distantly fixed on some fuzzy dream while walking a valley in which rogues regularly made huge fortunes—sometimes quite literally, as the thriving counterfeiting business illustrated. He was known by the blanket that he wrapped around him in the worst weather, an identifier that Summit County historian Lucius Bierce recognized in one of local poet Abner Robinson's attacks upon Kelley's uncompromising rigidity:

> Old Beelzebub when he gets him there
> Will take him by the throat
> And hold him in the brimstone fire
> And singe his blanket coat.[9]

Kelley's canal was the moonwalk of the early nineteenth century. Many people refused to believe that the thing could be done, even after the Erie Canal was completed. New York was, after all, a settled eastern state with big money and lots of farms, and the Erie's route was not nearly so topographically demanding as that being proposed in Ohio. The 308-mile Cleveland-to-Portsmouth canal would, if ever finished, require twenty-five miles of feeder canals, 1,260 feet of total rise and fall through

146 locks, sixteen aqueducts for conveyance over major streams, and 155 culverts.[10] Along the way, feeder gates would have to be placed to resupply the overtaxed waters, and sluice gates installed to relieve the opposite problem when floods came. An entire system of man-made lakes would also be needed to help overcome the basic problem of getting water to go where it doesn't want to, lakes that would one day wear vacation names like Portage and Buckeye and (on the sister Miami-Erie Canal) Indian. At Summit Lake, between Akron's high ground and Lock One in the descent toward Cleveland, the lack of firm perimeter ground for a towpath forced engineers to build a great "floating bridge" across the southeast part of the lake, an improvisation that in later decades would see mules going ankle-deep in the water seeping through the rotting boards. All of this had to be accomplished with engines no bigger than the muscles of man and animal.

The locks, themselves, were engineering masterpieces. Ancient in principle, these passageways to higher ground required thirteen-foot walls (nine for the lift, four to match the canal depth), huge gates of white oak that could be opened by a single man pushing a balance beam. Sluice gates were built into the lower part of the gates, and miter sills provided a watertight seal when the gates were closed. Into the locks pressed the demanding frames of eighty-five-foot packets and sixty-ton freighters. Then ex-President John Quincy Adams, passing through the valley on a packet in 1843, complained of "the careless and unskillful steering of the boat in and through the locks…The boat scarcely escapes a heavy thump on entering every one of them. She strikes and grazes against their sides, and staggers along like a stumbling nag."[11] It is probable that the old aristocrat did not fully appreciate the difficulty of the task.

There were forty-four of these great inland Gibraltars between Akron and Cleveland alone, each helping to break the precipitous 395-foot drop from the nation's watershed in Summit County to her north coast on Lake Erie. During nearly two centuries they have had to endure much more than the heavy bumps of roughly managed boats. Flooding, desertion, and the suitability of their magnificent sandstone blocks for other projects reduced them, where they continued to exist at all, to forgotten chunks of history left on unwanted land. For a little while in 1988, I thought I had found the remains of old Lock #43 down in the Flats of the Cuyahoga in Cleveland when I came across two sandstone block walls that had been used to support a newer bridge over the railroad track line. My hopeful reasoning was that: (1) the walls looked much like those of Red Lock; (2) this was right about where #43 would have been; and (3) the railroad had exactly replaced this part of the canal in the middle of the nineteenth century when Cleveland had wearied of it and moved the canal terminus three miles south. (By 1988 the city had, in a turn of poetic justice, wearied of and abandoned the railroad as well.) But it was not to be. The county engineer shook his head impatiently at my repeated, leading questions, pointing out that the old abutments were not nearly old enough to be old #43. Fleetingly, I then wondered if this newer construction had made use of the sandstones from one of the other locks. Probably my wish was father to the thought all along, so much did I want to resurrect another of these works that so excelled in both art and craft.

Even after completion, the engineering challenges remained. Because the canal walls were easily breached, the state had to keep emergency boats ready for dashes to trouble spots that could ground all traffic in a short time. Long before orange barrels, state transportation crews had learned the art of maintenance and repair. A strict speed limit of 4 mph was maintained because anything more than that would create a wash that could break down the banks. Spring and summer droughts threatened the sufficiency of water levels for heavy traffic, and winter, especially in the Western Reserve, naturally iced everything to a stop.

~~~

The chemistry of the canal was captured, then forever lost, in the unique lifestyle that accommodated her strange physical features. Ohioans would find many other ways to conduct the transporting business of their lives via trains, trucks and planes, but never again would they do so in quite so colorful a manner. Here, I think, is where the ghost is most restless, calling our furiously paced, egocentric world to take some time, even a moment, to remember the men and women who lived on the dirty ribbon of water, whose sons played hockey on its icy surface in winter, and whose daughters uncomplainingly washed the family clothes in it during warmer months. Two photographs define the difficulty of that remembrance. One displays the clean quaintness of the modern canal boat replicas, the St. Helena III at Canal Fulton or perhaps the Monticello III at Coshocton, filled with pleasant looking tourists wearing shorts and dangling cameras while a hoggee (the mule driver, perhaps a college student working a summer job) drives the placid mules on the towpath. The second, taken a century earlier, captures several people on a family freighter tending a clothesline that stretches across an open hold containing the refuse from a recent run. One of the daughters is doing the wash cycle in the canal beside the boat. The ragamuffin children and hard-looking father seem intrigued by the camera's interruption, but otherwise look to be in the middle of a typical day of life on a canal boat. Up front, two uneven, tattered shutters signal the cramped, year-round living quarters of the captain and his family.[12]

In a way, those two pictures reflect the alpha and omega of the lifestyles floating on the canal. In the timeless tension between rich and poor, haves and have-nots, the canal was a means for the former, an end for the latter. For ex-presidents like Adams, the canal meant getting you somewhere in good time. Their passages through Red Lock were filled with physical irritations and another glance at the watch; there was little thought about who might live up at the top of that hill behind the woods. But for those on the family boats, the canal *was* their life. When Mother Nye of a famous canal family by that name produced eighteen children, nine boys and nine girls, it simply meant that the family would have to run two boats rather than one. They were wedded to the great ditch itself, not to what lay at her termini.

The sleek eighty-plus-foot packets captured the eye if not the soul of the young, transportation-crazed nation. Built for speed these cutters took full advantage of the 4 mph speed limit, advertising the entire 308-mile Cleveland-to-Portsmouth trip in a stunning eighty hours (a claim that would require a stretching of either the speed limit or the truth, since each of the locks alone required approximately twenty minutes passage). One had to have experienced the alternative to appreciate this stupendous advance in comfort and speed. A foreign stagecoach passenger called our section of what is now Riverview Road the worst road he had seen in America.[13] It was not uncommon for passengers in these springless torture chambers to arrive at hostelries with cuts, bruises, and even broken bones. Boys were kept on hand to meet the coaches and attend to travelers' wounds, much the same as modern bellhops look to the luggage.

On the packet boats, mule or horse teams were exchanged at stables along the way, while the large crew of four to nine men saw to the tight efficiency of carrying seventy-five to one hundred passengers. The crew's quarters were up front, the kitchen aft, and the large thirty-five to forty-five foot center cabin provided a tavern by day and the men's dormitory at night. The business of turning all available space into sleeping quarters was an engineering feat in itself, with privacy and comfort far down the list of canal boat priorities. There is evidence that some of the packets had beds folded up against the walls that could be suspended outward at night. *Cleveland Plain Dealer* columnist, Georgiana Worthley, reported that the food wasn't bad—"salmon, shad, steaks, ham, liver, potatoes, pudding, bread, butter, tea and coffee"—but the smell from the kitchen probably was.[14] Too, in all my reading about canal boats, I never saw a design that hinted at a bathroom. Discreetly placed chamber pots and dashes ashore must have sufficed. But, at four cents per passenger per mile, this was ocean liner luxury through the Ohio hinterlands.

The people who lived closer to the ground traveled the canal at a cent and a half a mile, and a mile and a half an hour. These were the folks who would drop on board a freighter from a lock wall or an over-arching bridge. They sat or stood where they could, on top of long stashes of lumber or amid casks of produce, and made other arrangements for sleeping, the onboard space being pushed to its limits by crew, cargo, and the animal teams that were brought aboard at night and housed in a center stable.

For these passengers, the hours alternated between tedium and activity. In the mornings they could watch the quick-stepped ballet of the smart mules who knew—when many of the horses did not—that starting sixty to eighty tons of dead weight in the water required short steps, not long lunges. If their boat was heading against the gentle current they could observe the ritual of the down-current boat slacking its lines as it yielded to them in passing. The lock passages, of course, were always worth the watch. Sometimes, when two boats approached the locks at the same time, and the crews were a bit testy from last night's hangover or a late running schedule, they would fight each other for the right of first passage. In times of lighter humor, they left that determination to the hoggees who would duel on the lock walls with their driver sticks. The first one knocked into the canal was the last one through the lock. The hoggees were, themselves, worth a second look. These were ten and eleven-year-old boys who led the canal parades at the front end of two-hundred-foot tow ropes because they were too small to handle onboard tasks. They were younger sons or brothers on family boats, or runaway waifs on company boats, a tough pack of Huck Finns cussing and smoking and fighting their ways to early manhood, their tireless feet thudding out the heartbeat of America's growing economic might.

At the apex of the canal's social structure was the captain who by definition not only ran the boat but often owned it. Canal purists insist on both of these attributes, and their punctiliousness is probably in order since the dual role implied both prosperity and authority. These were perhaps the most colorful of all canal folk, and have managed to generate a separate body of folklore. They competed with each other not only for the business they hauled but also for scarce supplies of space and attention. Sometimes the canal just wasn't big enough for more than one such ego at a time, and it took a crew-to-crew brawl to sort things out. The captains were known by their outlandish dress, that sometimes included a stovepipe hat and at least semi-military apparel. Captain Brad Voshall described himself as a "peaceable man" who once shot a fellow canaller only because "I happened to have a gun," but who otherwise preferred his fists. In later life, Voshall was notorious for turning a dinner invitation into a two-or three-week stay, during which the half-crazed host(ages) would be treated to all of the canal stories in the captain's repertoire.[15] After Chingo Smith's fictional Captain Lumm had cleaned house of several detractors of the Erie Canal, which he was obliged to do one-handed having damaged his left on a stagecoach ride, the tavern wench complimented him on his great fighting ability. Lumm brushed aside the remark: "Shoo! We wouldn't call that fighting on the canal," he said. "Just sparring. When canallers *fight*, someone most genrally gets hurt."[16]

With the passing of the canal's heyday in the middle of the nineteenth century, the world no longer had much of a place for these men. In dozens of taverns dotting the valley they would, for a bought round, resurrect the well-screened memories, slowing their verbal pace only long enough to wince at the sound of the freight train whistling by outside. Still later, in rest homes, they would babble their worn-out tales to the walls and unhearing nurses. No one seemed to realize, or care, that when their voices were finally silenced so, too, was the voice of the canal.

Many of the captains were still alive when the end came in 1913. Mercifully, the canal's fatal blow was swift and complete, sparing their watery, old eyes the continuing agony of any lengthy death throes. The captains' hopes had risen a bit when some lawmaker in Columbus had pushed through the legislature a public improvements bill aimed at rehabilitating some of the badly weathered concrete work on the lock walls, but it proved a futile gesture. By 1913, the canal was finished. Large sections were no longer navigable and, save for church picnics and an occasional, enterprising grocery boat, traffic had disappeared. Better to end it with one tremendous blow to the whole body than to see it dismembered piecemeal, with each lock's demise like an old friend being hauled off for burial.

It was Mother Nature, not man's hand, that delivered the blow. With bittersweet irony she had decided that, with all of the elements available to her, she would do it with water. For the occasion, she brewed up the greatest flood in Ohio's history. Unlike the more publicized but localized 1937 flood, this one would reach virtually every part of the state, as if determined that not only the entire length of the Ohio & Erie but all of Ohio's other canals would fall under one heavy sweep of the boiling waters.

Perhaps 1913 was the perfect time for it. That year was the last day of an Indian summer in a world heading toward a brutal winter. The bright optimism of the late nineteenth century still flickered in millions of hearts conditioned to believe that the flow of history was positive and progressive, that the spread of Western culture throughout the world—even on the horns of blatant imperialism—was ushering in a late paradise. William Jennings Bryan said as much to 1,500 Akronites days before the flood, assuring his audience that the progress made toward literacy and temperance meant that the world was growing better.[17]

Like a thirteen-year-old boy who simultaneously gives the appearance of going forward to manhood and backward to boyhood, 1913 seemed to have one foot on either side of a historical time line. The *Akron Beacon Journal* edition that reported Bryan's speech also carried the story of the death of J.P. Morgan. No more would America's economy be in the hip pocket of one man, no matter how rich he was. As if to underscore the point, the nation saw the passage of the first income tax law that year. A smaller article in the March 24 edition quietly noted that the New York

Fire Department had purchased its last horse. There was also news of an indoor baseball league, and a technologically mystifying dictaphone interview with actress Julia Dean in Washington, DC.

But the same newspaper also reflected much of the nineteenth century. The *Beacon Journal* thought it newsworthy that one of Akron's own women had asked her husband for a divorce, and the want-ads carried an item calling for a "chauffeur: light-colored." The comics section was embarrassingly ethnic in its humor (*Hans & Fritz, Mr. Batch, Their Only Child*), and a regular column was, *What Funny Creatures Men Are*. Occasionally, there were odd bits of "scientific" knowledge for the readers, especially the large number worried about their health. It was matter-of-factly noted that left-handed people do not make good public speakers, that doctors could help men with "weakness" ("Are you a sound, manly man?"), and that Wrigley's Gum benefited both digestion and teeth. Other products with familiar echoes included Listerine, Fruit of the Loom, Franco American soups, and Post Toasties. Dentists advertised gold crowns for $3.00, while banks offered $25.00 and $30.00 loans that could be paid back at sixty and seventy cents a week. For entertainment, the local arena offered wrestling, with Matsuda the Jap having a go at Ivan Dorsey. The songs people sang tended to look backward, "Hail, Hail, the Gang's All Here!" from Spanish-American War days, and "We'll Have a Hot Time in the Old Town Tonight," recalling the Chicago Fire.

Quietly, on the far eastern horizon, black war clouds gathered that would formally initiate one of history's most horrible centuries. When the firestorms of two wars and a depression had cleared little would be left of the world of 1913. It was a historical kindness that the canal was forever frozen in that last year of lingering, naive hope.

The beginning of the end was Easter Sunday, March 23. Three major air masses conspired to dump an unprecedented amount of rainfall on land that had already reached the saturation level under the soggy prodding of late winter snow melts and rains. From the western plains came a savage storm front that spawned tornadoes in Omaha and Terre Haute. In what became a rather bad habit of sensationalizing over the next few days, the *Beacon Journal* at first proclaimed 500 dead in Omaha before dropping the figure to 225 the next day (March 25). But the newspaper missed another ominous message from Omaha, only casually noting the impact of the pouring rains on a men's meeting in town. The second and strangest of the currents began far to the northeast and displayed its nor'easter temper by fighting its way westward up the St. Lawrence, over Lake Erie, and toward northwestern Ohio. From the Gulf came a warm air mass, friendly enough by itself but disastrous in combination with the other two rogues with which it collided somewhere over the middle of the Indiana-Ohio line.[18]

Up on top of Red Lock Hill in 1913, the land that would become ours forty years later was part of a farm, probably owned by a man named Hazanaugh, or perhaps Gainey. Little hints, like the rusted pieces of barbed wire that have eaten themselves halfway into large maple trees, lead me to believe that it may have been pasture land at the time. While I can't be certain of that, I can make a fairly good guess what it sounded like there on Tuesday, March 25, 1913. The roar of our little waterfalls, always noisy after heavy rains, might have drowned out the Wagnerian Valkyries that night. The *Beacon Journal* subsequently reported that 9.65 inches of rain fell on the already soaked ground between Sunday and Thursday, with nearly five of those falling between Monday and Tuesday mornings.

The falls at spring rush. The picturesque water fall behind our property could turn from a trickle to a torrent when heavy rains came, hinting at what the larger valley must have been like during the canal-killing flood of 1913, Ohio's worst ever. In gentler moments, the falls bear a close family resemblance to nearby, oft-photographed Blue Hen Falls. But our falls must have been the shy sister. *Photo from author's collection.*

Rivers, streams, and lakes forgot how to behave. So, too, did the canal. The Cuyahoga Valley, especially during that long Monday night, called forth the vestige of ancient Lake Cuyahoga from its glacial bed, pulling the waters fourteen feet up its walls in the Pinery Narrows just north of the Rt. 82 Bridge. At some points, the Cuyahoga took on the dimensions of the Mississippi, a solid mile across, and the huge body of water was reported to be thundering down the valley toward Cleveland. The Valley Railroad (later the B & O, and then Chessie) was reported to be under six feet of mud in places, and eight homes along the canal in Boston slipped into the savage waters. Normally identifiable streams such as Tinkers Creek and the Chippewa River twisted into the boiling waters of the Cuyahoga and the canal to create one raging torrent. For eighty-three years the canal banks had known only slack water, but on this last night of her life the waters would pour through and over her at fifteen miles per hour.

Some people believed it was the end of the world. The *Beacon Journal* helped that idea along a good bit. When the telegraph communications to the outside world were severed, the newspaper became the conduit for the worst of the flood's rumors. The Wednesday morning edition reported that 5,000 people may have died in Dayton—the actual number was a single digit percent of that amount—and that unconfirmed reports from that city indicated that men were shooting their families, then themselves, when they saw no escape from the rising waters. But the *Beacon Journal* may not have overstated the case when it pronounced the long Monday night the most strenuous twelve hours in Akron's history. That day's edition was a single page, printed by the power of motorcycles used to run the typesetting machines.

The canal-killing flood of 1913. Akron. The death knell for the Ohio & Erie Canal was sounded during the statewide flood of April, 1913. The magnificent locks, in particular, suffered fatal damage from the thunderous waters. In some cases, as in Akron, locks were deliberately dynamited when debris gorged them into instant dams, severely compounding the flooding. Needless to say, the acts were not appreciated further along the four hundred foot downward slope below the Akron summit. *Photo provided by the Summit County Historical Society of Akron Ohio.*

All over the state, people's primary response to the disaster was to try to get out of harm's way. In Cleveland, a huge freighter broke loose of its moorings and smashed a bridge over the Cuyahoga. To the south, in Columbus, the waters of the Scioto reached the second stories of West Side houses, where they would leave a discernible bathtub ring for two decades. The Mad River, far to the west, re-earned its name by claiming a train. And, in hard-hit Dayton, a large number of people were roof-hopping and rope-sliding between buildings, acting out a desperate irony in fleeing not only the rising waters but also the more dangerous fires that ravaged several downtown blocks when the gas mains were not turned off in time. During the entirety of those horrifying forty-eight hours in Ohio only one significant attempt was made to do anything about the flood waters themselves. That attempt put the last nail in the coffin of the Ohio & Erie Canal.

The main problem was those magnificent locks. In Akron, when loose lumber and other debris began choking the south gates of the locks, the water that backed up behind them threatened the city with serious flooding. Virtually every business in the city along the canal—no small part of Akron's economy—was in danger. One building had already dropped into the formerly friendly water. The problem was further exacerbated by the swelling waters of the Portage Lakes on the high ground above the city.

At some point during the night, somebody began dynamiting the gorged locks. The confused, early newspaper accounts did not know whether to portray the acts as heroism or vandalism, but even the police seemed to approve of this logical necessity. (The logic was lost on people down in the valley when they realized that their neighbors had just sent them some more water they didn't need at the moment.) Subsequent time would prove that city officials themselves had been involved in the destruction of the locks.

But survival was at stake. If anyone shed a tear for the Ohio & Erie Canal that night it was forever lost in the thrashing floodwaters.

∼∼∼

Maybe the ghost will yet be laid to rest. In the 1970s, much of the Cuyahoga Valley in our area was purchased by the federal government as part of the burgeoning Cuyahoga Valley National Recreation Area, now the Cuyahoga Valley National Park (CVNP). The faint depression of the lost canal bed that Dad and I walked that spring day is now part of a 110-mile canal tow path restoration project between Cleveland and New Philadelphia. Today's earthmovers easily plow out the former byway to the whispering sighs of hundreds of Irish and German spirits lying in nearby cemeteries. A smoothly paved hiking trail tops parts of the old tow path formerly known only to mules' hooves and hoggees' feet. As many as three million people annually visit the CVNP grounds once known as The Black Forest.

Occasionally, a semi-serious voice is heard advocating the construction of a new Ohio canal, interestingly for the same reason the first canal was needed: the high costs of overland transportation. But it probably won't happen, and it wouldn't be the same anyhow. The construction costs would be in the billions, not hundreds of thousands; boats would be self-propelled and could do far better than four miles per hour; and pumps and piping would probably make reservoirs and even locks unnecessary. The old canal, the great canal, was a once-in-forever proposition. I am grateful to have been on her right-of-way.

Notes

1. William D. Ellis, *The Cuyahoga* (Dayton, Ohio: Landfall Press Inc., 1985), 108.

2. William I. Barnholth, George Croghan, *Cuyahoga Valley Indian Trader* (Northampton, Ohio: the Northampton Historical Society, undated), 15.

3. James S. and Margot Jackson, *The Colorful Era of the Ohio Canal* (Akron: the Summit County Historical Society, 1981), 21.

4. Samuel Hopkins Adams, *Chingo Smith of the Erie Canal* (New York: Random House, Inc., 1958), 75.

5. William D. Ellis, *The Cuyahoga*, 109.

6. James S. and Margot Jackson, *The Colorful Era of the Ohio Canal*, 11.

7. William D. Ellis, *The Cuyahoga*, 113.

8. James S. and Margot Jackson, *The Colorful Era of the Ohio Canal*, 11.

9. General L.V. Bierce, *Historical Reminiscences of Summit County* (Akron, Ohio: T.&H.G. Publishers, 1854), 47.

10. Harry Scheiber, *The Ohio Canals* (Athens, Ohio: Ohio University Press, 1969), 43.

11. James S. and Margot Jackson, *The Colorful Era of the Ohio Canal*, 18.

12. Jack Gieck, *A Photo Album of Ohio's Canal Era, 1825-1913* (Kent, Ohio: The Kent State University Press, 1988), 240-241.

13. James S. and Margot Jackson, *The Colorful Era of the Ohio Canal*, 9.

14. *Cleveland Plain Dealer Magazine*, May 22, 1932.

15. Terry K. Woods, *Twenty Five Miles to Nowhere* (Coshocton, Ohio: Roscoe Village Foundation, 1978), 26.

16. Samuel Hopkins Adams, *Chingo Smith of the Erie Canal*, 35-36.

17. *Akron Beacon Journal*, March 13, 1913.

18. Alan W. Eckert, *Time of Terror* (Dayton, Ohio: Landfall Press, 1981), 3-4.

3
Altered Altars
The Buildings

Jonathan Hale's home. Bath, Ohio. The valley's earliest mansion.
Photo by Ian Adams.

Some notable structures of the Western Reserve • Perils of home-building in the 1950s • Of School and schools • The foundry that became a shopping strip

The structures of our lives are the perishable parts of our past. The land stays forever—buckling, submerging, foresting, but always there. The people come and go, but they have clever ways of perpetuating themselves in words, events, and the leavings of their lives. But the vast majority of our dwellings, be they for business or pleasure, are terminal from their first brick, block, or board. They will last only as long as they serve some useful purpose, or until someone comes along with a bright idea that requires some of the world's most finite commodity—space. There's the rub. Discarded human bodies can be disposed of at little inconvenience to the rest of society, but not so the edifices that housed them when they breathed life. A house will usually outlive its builder, but seldom its usefulness.

Buildings are almost always seen that way, strictly in light of their service to men and women. Occasionally, one manages to scale the heights of great art or significant history such that we allow it to stand alone, on its own merit. But these are the rare exceptions. We generally proceed on the assumption that people make buildings, not the other way around.

Sometimes, I wonder.

The Cuyahoga Valley has had its share of notable buildings. It is momentarily tempting to take the tour guide approach of listing the headliners, say, the Terminal Tower in Cleveland, or the Quaker Oats Building in Akron, not to individually mention the dozens of universities, sports arenas, and museums. Millions of good stories are soaked into such monumental structures, and few in the Western Reserve have managed to avoid all of these. But this is not where most of us really lived. If you want to know about them, take the tour bus.

Closer to the mark are the smaller, more lastingly substantive structures that have carried with them important stories over greater periods of time. Wilson's Mill (or, its former moniker, Alexander's, pick your era) still sits beside Old Lock #37 along the extant section of the canal on the way to Cleveland. As late as 1969 the mill was using the water from the lock spillage to help drive the waterwheel and the business that the family still operates as a feed and grain store on the historic site. A stone's throw away along Canal Road are the Hynton House, probably built around 1809, long before its more famous Hale House neighbor in Bath (with which it shared a common architect), and a smaller stone house at Stone Road built into a hill and over a natural spring, probably in the early 1820s.[1] Farther south, in our part of the valley, sit several look-alike houses in the town of Jaite, an old paper mill company town bisected by the then B&O Railroad line across the river from Red Lock Hill. With nearly all of the twentieth century's years under their belts, these company houses witnessed the death throes of the canal, the company, and the railroad, and withstood the frequent floodwaters of that grand river that enticed boat and steam engine to the valley. The houses hit a bit of good luck in what otherwise would be their declining years, and are now nicely renovated as the administrative offices for the Cuyahoga Valley National Park.

Other buildings have left only the outlined traces of their foundations to hint at their glory days. Somewhere in the tangled underbrush at the foot of Pine Hill, below the Rt. 82 bridge that spans the river's Pinery Narrows, are the scattered pieces of a Civil War ammunition dump. The research on it is as tenuous as the remains, but Joe Jesensky, who knew the valley's story about as well as anyone, identifies the site as Crazy Man's Hollow, a name derived from the Civil War veteran who slowly lost his mind while on assignment at the lonely post. Some say the veteran filled his walls with bizarre formulae and drawings of mechanical contraptions not of his century.[2] Meanwhile, at Brandywine Falls, which now hosts a fashionable bed-and-breakfast, stone foundational remnants are all that testify to several structures that lived off of the power of the plunging waters: a distillery, sawmill, and woolen factory. As I said, traces. Architectural hieroglyphics possessing old secrets, proud verticals now reduced to embarrassed horizontals; scratches in the land's everlasting reclamation project.

Yet, as kids, we were much intrigued by ruins, probably because they teased our imaginations into overdrive. A perfectly formed and maintained structure neither entertained questions nor admitted of possibilities, but a collapsing wreck was a half-finished novel. Deep in the woods we once stumbled upon a small, weatherworn house with the windows and doors all boarded up and covered with some faded warning signs. Mom or Dad told us to stay away from it, that the place had been fumigated then apparently abandoned. But my mind never abandoned it. I still wonder about the inside of that house. What kind of creatures lay dead on the floor? Wouldn't it have been a great place for robbers to hide out? Did even a single shaft of sunlight ever penetrate its murky darkness?

Somewhere else in those same woods I once saw a small coop of some sort, probably for chickens. I didn't explore that one, either, just gazed at it in wonderment, tantalized by the prospect of this little bit of cozy order in the natural chaos of the woods. For the remainder of my childhood after that one sighting I looked for that coop, like a dream-walker in search of a vision that was there only a moment ago, hungering to explore its tiny space, wondering endlessly why such a thing would be built in the middle of the woods, and haunted by its disappearance. I never found it. It was probably knocked down and cleared away during one hour of some man's chore-filled morning, nothing more than a thought finally fulfilled: "I better get that old shed out of there today; I've been putting it off for years." But he couldn't knock it out of my mind. There, it had been built to last.

Perhaps that explains our juvenile fantasy affair with building forts and cabins. These structures, seldom more than a couple of old sheets tacked to a tree or some scrap lumber nailed together, were palaces in the unencumbered architecture of our childhood imaginations. Anything we could crawl into qualified as a success, whether or not the rains were very much slowed by what was overhead. Our most memorable effort was a fort we constructed in the uncleared portion of Uncle Frank's front yard. In addition to the standard tent-like sections, it also included a "room" actually made of boards. The room was probably not more than eighteen inches wide, and you could only crawl in far enough to say that you had been there before having to exit the same way using reverse gear, but to us there was only a difference in degrees between it and our real bedrooms. As we weren't quite ready to tackle plumbing, we simply dug a little ditch outside the fort and called it a "leak bed," figuring that nomenclature was close enough to the leach bed Dad had to dig behind our house a year or two earlier. Cousin Dave, older than Lloyd, Jack, and I, and always ready to go one up on us, threw himself into the construction of a fort with a basement. He dug a large hole not far from our effort, but soon thereafter abandoned it to the frogs that gratefully filled it with their presence and songs each spring for many years to come. One thing David did complete, however, and that drew our unlimited admiration, was the mounting of an old faucet into the side of a drum such that he could produce running water with a turn of the handle. We could only shake our heads and marvel at that miracle.

~~~

Jonathan Hale built the Cuyahoga Valley's best known and preserved homestead. Both the man and the means behind the Hale House story are appropriate reflections of the early development of the Western Reserve. Here was no Alfred Kelley, builder of canals and railroads, or Samuel Huntington, seaboard political transplant who would gather up the governor's office and several others along his ambitious way, or even David Hudson, the wilderness visionary. Hale shared with these men only the same westward trails they took from the east to the three-million-acre dream on Lake Erie's southern shore known as the Western Reserve. He was a large, raw-boned man with an enormous capacity for work and heartbreak and hope, yet he could also be found drifting around the valley directing church choirs and taking occasional stabs at poetry. With apologies for the anachronism, there was always something of the literary Charles Ingalls in him, a quietly talented and deeply religious man who chose to plow his human gifts into the ground at his feet rather than ride them to peaks of what is usually thought of as achievement. Even Old

Brick, his historic three-story mansion that set a precedent for dwellings in our part of the valley, was pragmatic in nature, expanding with incremental spurts of construction as need dictated, and begun only after years of the seven-member family living in a former squatter's cabin.[3] Hale House was less a monument to architecture than a reflection of Hale's practical arrangement with the new land. The house was big because Hale had a big family. Every cubic inch teemed with life.

Many might think that the fabulous homes along Euclid Avenue were better testimonies to the Western Reserve's story. Author Peter Jedick hints that during the latter half of the nineteenth century many people considered that avenue to be the most beautiful street in the world, easily rivaling the Prospect Nevsky in St. Petersburg and the Champs Elysees in Paris. Home to the likes of John D. Rockefeller and Senator Henry Payne, and host to a glittering stream of renowned visitors with names like Grant, McKinley, Taft, Carnegie, and Morgan, Euclid Avenue was a freeze-frame of the seemingly unlimited possibilities of America's Gilded Age. The unimagined personal wealth that spread itself ostentatiously under the avenue's stately elms screamed the Reserve's success story to the world. The 80+-room-mansion of crude oil refiner Samuel Andrews, three years in the making, including castle towers and turrets, was built in the confident expectation that Queen Victoria would see fit to christen the place with a visit. It was occupied only three years, after which it sat vacant but fully furnished (the palace simply couldn't be managed as a home) for the next thirty years, then demolished. Much of the valley's best and worst was woven into the golden brick that paved Euclid Avenue.[4]

Big money can be its own worst enemy, not infrequently because it encourages similar greed. Some lesser entrepreneur managed to squeeze a dry goods store on a downtown corner of Euclid Avenue, and the seeds of destruction began to germinate. Andrews' castle was headed toward a future as a miniature golf course, while Rockefeller's home was ironically destined to be knocked down in order to make room for a gas station. Euclid Avenue did not last because the transitory values of unlimited and unconscionable capitalism did not—could not—last. But Jonathan Hale's house was built of sterner stuff. There is at least a piece of the Hale House story—and virtually none of the Euclid Avenue story—that has been reenacted in thousands of the homes built in the Cuyahoga Valley. In some ways, our home in Northfield was an heir to that legacy. Jonathan Hale assumed a huge risk in undertaking Old Brick. On the plus side, he had a moderate degree of prosperity, good standing as a landowner, and the news that the proposed Ohio & Erie Canal would plow right along his property line. But prosperity in early valley days was a tenuous thing, hard money was scarce, and the prospects for the canal were still seen as somewhere between boon and boondoggle. For a long time, the house was only a whisper away from being "Hale's Folly." Mom and Dad took an even greater risk when they decided to build "out in the country" in 1953. There was no squatter's cabin or large chunk of property to fall back on if the thing failed.

Too, our home was always inextricably linked to the land. Not in the strict economic sense of the Hales, of course, where limestone was quarried for sale and ground clay used for brick making. But where Hale sold casks of lime to the infamous John Brown's father in Hudson, we sold black raspberries to Virginia Smith and Emma Copen. And the humble ranch style box we dropped unobtrusively onto the plot of ground at 846 Glencrest Drive was never as small as the walls. Its boundaries included a ball field, waterfall, a woods, and a stand of crabapple trees. It was a package deal.

Times being what they were in 1953, our family was not able to enjoy the luxury of moving into our new home all at once. Instead, we did it in stages, advantaging ourselves of time, money, and the aid of others as they occasionally became available. To my knowledge, the only formal training my dad ever received was as a sonar repairman during World War II, while Mom used those same three years to gain her professional training as an x-ray technician. Neither of those experiences was likely to be of much help in building a house. Yet my parents assumed that burden as readily as I now might buy a car or plan a vacation. Hardship and delayed gratification were accepted parts of the American Dream at the time.

A few hundred borrowed dollars for a lumber kit and a how-to book were all that Dad had to get started on the house. I'm not sure how much time, if any, he spent worrying about the consequences if this little project blew up in his face. With their limited resources and vanishing opportunities in the recession-bound, post-war economy, such a failure would, in all probability, have buried them for good. The financial world has often been less forgiving of other little people for smaller mistakes. Today, the idea seems so ill advised, risky, primitive. But today is the world of wedding-present homes and starter-mansions; Mom and Dad's "today" was 1953. Into that year they carried two broken homes, the modern world's worst depression, history's most terrible war, a long string of hospital bills from Mom's emotional troubles of 1948, and the same flickering hopes that have teased twenty-somethings for a hundred years—maybe a thousand. These poorly grounded hopes were not to disappoint them; Mom and Dad would live in that same house for the next fifty-five years.

The big home-owning gamble would have never had a chance had it not been for Uncle Frank. Five years older than Dad, Uncle Frank, with

> Ground-breaking, 1952, Sagamore Hills. Dad and Uncle Frank (far background) dug the footer by hand. For me (near background), the hard work provided a lark which, I assumed, was a part of every kid's childhood.
> *Photo from author's collection.*

college degree in hand, had been the first of the Knowles boys to escape Cleveland and carve out a toehold in what was then known simply as "the country." Somehow, he had been able to grab a piece of land in Northfield, fifteen miles to the south, after which he set about the business of settling on it. Here, things got a bit tricky. He could not, of course, get a home loan without significant collateral, and he could not hope to show such collateral without a home. Therefore, he hit upon a plan that was both brilliant and in keeping with his sense of humor. He decided to begin building a house, or at least something that qualified as a structure, stop at some point when the thing approximated an identifiable geometric shape, and then proceed to the bank for a home improvement loan. To complete even this shack he was forced to scavenge for pieces of lumber that he would nonchalantly haul aboard the city bus on his ride home. I assume he did all of this with a reasonably straight face, although the same assumption cannot be made for his fellow bus passengers or the bank manager. At any rate, it worked. By the time his little brother was ready to make the move from the city Frank Knowles was ready and willing to offer him a couple of his acres next door to his glorified box at 858 Glencrest Drive.

Thus began the phasing in of our new house. Decades later, I still remember more about what wasn't finished than what was. My four-year-old eyes saved for me images of a shovel-dug footer, hand-mixed cement, scraps of tar-paper embedded in patches of mud, and someone's arbitrary decision to "leave that little maple there for the front yard." That scrawny two-foot sapling would begin its slow pursuit of a shadowy fifty-foot destiny as a green canopy for the Lorence Cook Knowles family.

I also remember the unfinished floor. More than any other part of that house, it was the floor that first gave notice to the surrounding wilderness that there would henceforth be some order here. The shaggy black locusts and baggy crabapples looked down upon their well-milled cousins lying in neat rows under the gleaming sun and grudgingly moved over to concede these city folks their silly, little dream. On this postage stamp of symmetry in the midst of nature's vast chaos, I would lie on my back and stare up at a sky full of stars such as I never suspected existed above the perpetual dull glow of Cleveland's lights. The walls—well, they might come tomorrow or the next day, and the roof was too far in the future to think about. For the moment, the floor was everything. Next to me a bulky radio, umbilicalled by extension cords to some unseen outlet in Uncle Frank's house, sent the corny sounds of *One Man's Family* into the chill of the night, the title serving as a fitting testimony to the work at hand. I also liked *Fibber McGee and Molly*, though I can't remember why.

The exposed flooring supporting my lazy observance of God's astronomical splendor was never to see a life of quiet comfort. When Lloyd and I became rambunctious, particularly during the eleven-to-thirteen-year-old days, when the value of everything in the house was measured by whether we could reach it with a jump, Dad would occasionally bring out the ultimate warning, "You guys knock it off! Remember these are single floors." In high school, by which time I could touch my elbow on the ceiling (though I don't think Dad knew about that accomplishment), I had visions of ending my landing thigh deep in splintered floorboards. But the floor's first adversary was my five-year-old cousin, Jack. In racing across the partially finished surface one evening, Jack failed to distinguish the subtle color difference between the floorboards and the insulation cover. He went in up to his armpits. After the obligatory questions about Jack's health, there was some tight-cheeked grumbling from the men who would have to make this mess right in the morning. I don't think any of it was heard by Jack, who, smiling broadly, assumed that the trouble had ended with his safe airlift out of the itchy pool. But for many years to come, I suspect that my meticulous dad was feeling around the living room floor with his bare feet, looking for that cold spot.

We could not, of course, live on bare flooring exposed to the sky. I don't remember much about the transition from our second story flat in a rickety Cleveland rental to the new homestead, but there must have been many trips back and forth before we were ready to make the leap to the new lillypad. One of these provided me with my first experience with injustice and anger. It happened late one warm spring evening, probably after a long day of labor on the house and with the prospect of a late night ride back into Cleveland yet ahead. Mom and Dad were enjoying a few companionable moments around an outdoor fire with Uncle Frank, Aunt Doris, and, I am guessing, Janet and David, my two older cousins. Lloyd and I had been banished to the car where our orders were to "Go to sleep; it's late." The problem was that the car, parked in Uncle Frank's drive, was no more than thirty feet from the festivities. Even under the best of circumstances four-year-olds don't like to go to bed, mostly because they (rightly) suspect that they will be missing something. In this case the evidence was right in front of my eyes, now filled with stinging tears. There may have been greater injustices somewhere in the world, but at the time, I could not imagine any. Lloyd's sense of judicial propriety served him better; he fell asleep.

But there was nothing cruel about putting your kids to bed in the family car. Cars meant something different in the 1950s than they do today. In some ways, they were as big a part of the American Dream as were homes and land. Taking an hour drive was a perfectly legitimate form of entertainment, even when you weren't going anywhere in particular. "Let's go for a drive" was the mid-twentieth century equivalent of "Let's go to the mall." Most of the boys returned from World War II to jobs that they reached by foot, bus, or train. All of them could tell you the exact hour they became owners of a family car. The car was a four-wheeled extension of the home, complete with heat, light, entertainment, storage compartments, chairs, and windows. Small wonder our parents were forever having to tell us that we couldn't play in this marvelous, miniature home. But even for the adults there was some irresistible, initial attraction that has long since been lost, probably the sense that a man who never before could claim more than a few paltry cubic feet of rented space to his name suddenly could assume ownership, if only for a passing second, over car-sized chunks of America dozens or hundreds of miles from home. Here was ownership you could take with you.

I have heard many Baby Boomers say that among their fondest memories is that of falling asleep in the back seat of the family car while riding home late at night. Perhaps it was the hypnotic drone of the tires, or perhaps the secure knowledge that their dad was physically in control of this moving box, but there *was* something special about falling asleep that way. Many of those same people will tell you that the beginning of the end of their innocence was not a dirty magazine or some deliberate cruelty, but rather the first time they heard their parents say, upon arriving home after a long, nocturnal drive, "Get up, we're home…No, you're too heavy to carry anymore."

Since my brother was less than two years older than I, we had some logistical difficulties with sleeping arrangements in the back seat once we outgrew the dogseat. The preferred real estate was, of course, the "inside" of the seat against the back cushion. Here, the art of car sleeping reached its highest form, gaining all of the advantages of comfort, security, and warmth. The guy on the front edge of the seat fared less favorably. In that position, the backward slope of the seat provided a dangerously false sense of security, especially when bleary-eyed dads broke the sleeping rules by making use of the brakes. A darting dog or a changing traffic light had a way of suddenly effecting a curve in your anatomy that looked remarkably similar to the back floor hump covering the crankshaft. Long before we worried about going through the windshield, we learned to fear that midnight catapult from the front edge of the seat into the oblivion below.

My particular problem was that Lloyd was rather good at angling for the inside position after he had supposedly gone to sleep. After dutifully begrudging me my turn on the inside, he would begin a slow wedging movement that would gain the favored position before the ride home was half done. When I complained or pushed back he simply continued to feign unconsciousness, a state in which he could avoid the moral consequences of his naked aggression.

It does not require a great deal of imagination to understand some rather immediate problems facing the family whose home is being completed in stages. The plumbing—I assume because it was so expensive—came later in the plan. Our necessary alternative became known as the "honey bucket." The honey bucket gave me my first steady job, that of flashlight holder for Dad who saw to the burying details each night. As silly as it sounds, there was a sense of ceremonial ritual to the procedure. Dad would carefully dig a pit deep enough to allow the newly cleared yard to maintain some dignity after the sordid affair was finished. The hole was always the same size, even in winter, when Dad's shovel served better as pick than spade. As the flashlight proved no truer than my wandering interest, there were two or three predictable admonitions from him to keep the light on the spot where he was digging. And, always, there was the little, unintended dance that he would do to tamp down the filled in dirt. Dad must have kept a pretty good mental record of the used spots, for I don't remember him ever hitting an old mine. Years later, I would grow some of Northfield's best cantaloupes in that very ground.

The honey bucket was not to be confused with the "slop bucket," the latter being reserved for used dishwater and spitting after teeth brushing. I might have forgotten about the slop bucket altogether had it not once become the center of a dispute about physical dexterity. Lloyd was absolutely sure he could clear the fetid pot with a standing broad jump. Being skeptical of that claim, and having nothing particular to lose in the prospect, I encouraged him to give it a try. In all fairness, it was a very good jump. He soared high, higher than was necessary, but unfortunately he had not made a similar provision for distance. The result was one of those clear images in my memory that lasts a lifetime: Lloyd, standing uncertainly knee-deep in the bucket, giggling, though not nearly so hard as I. Mom, who usually interpreted loud noises as threats to her children, came running into the room with panic in her eyes. It was probably relief more than humor that drew a laugh from her, for she could not have helped but see that there was an unfunny mess that would need cleaning up when the laughter died away. The funniest part was Lloyd's continuing quandary. Having become a part of the slop bucket, he sensed that he was not free to reassume his role as a human being at liberty to walk around the house. But staying put also seemed like a losing proposition, especially as the bucket was somewhat rounded on the bottom. He appeared to be groping for some kind of precedent that could provide a bit of guidance but, as there was none, he continued to just stand there, wobbling drunkenly like a potted tree in a strong wind. In his confusion, he failed to realize that the point about messing up the house was moot anyhow since his landing had already splashed most of the noisome water out onto the floor.

Dad took a lot of kidding over the years about his painstakingly slow pace in finishing the house. While we did not have to grow to manhood without indoor plumbing, there was always something that remained unfinished. It could certainly be said in his behalf that since he was not a professional builder, and was building the house with the little time left over from a demanding, full-time job, he could hardly have been expected to make good time. But neither of those was the issue. Dad was just careful—agonizingly careful. Most of the jobs that he did on that house only had to be done once, a fine tribute to his workmanship. Yet fifty-five years later, it could still be argued that the place was not quite finished. There was a room between the kitchen and family room that was variously called the utility room, carport (why, I don't know, since it could never have hoped to house a car), and backroom. I was never sure what to call it since it never took on any final form.

Ours was not the only house in Northfield that was slow to develop. Shortly after we moved in, Aunt Doris took Lloyd, Jack, and me on a hike through a large field just south of Highland Road a few hundred yards from our property. There, hidden by a choking patch of weeds and goldenrod, we came upon a rough foundation upon which rose only a single tier of concrete blocks. Aunt Doris explained that a man had started the house years before and had subsequently run out of money, but was planning to finish it one day. I believe the remains of that foundation still lie in that undeveloped field. The man is no longer likely to finish his house. He is surely dead. In some neatly kept cemetery, his tombstone probably reads, "Here lies a builder who was slower than Lorence Knowles."

~~~

It is one of those "take it to the grave" images embedded in my mind, as vivid now as it was in 1953: Russell Scholle's feet dancing hysterically six inches above the ground, and below the writhing crucifix his five year-old body. One hand was firmly anchored to his mother's, while the other was being pulled by the kindergarten teacher, Mrs. Steprow. As a rational adult, I know that scene could have only lasted a second or two at most; at the time, I might have said an hour. Then, his knees buckled and he was kneeling on the ground. His flustered mother pried his clutching fingers from her hand. Mrs. Steprow helped him to his feet and marshaled him to a seat. At that moment there were probably a hundred thousand hysterical Russell Sholles in untold numbers of American kindergartens across the land. And each of them was doing a variation on Russell's airy dance, emotionally suspended between the comfortable world of mom and the fearful world of school. It was here we learned that the birth canal was not to be life's only tough entrance.

But, generally speaking, I was not seeing things too rationally in September of 1953. What I saw was a little boy whose mother and teacher were trying to tear him in two. To my horrified eyes, Russell Scholle had been suspended like that for an eternity. What kind of torture chamber was this, anyway? And why was his mother acting as a willing accomplice to this hideous conspiracy? For that matter, wasn't my mother in effect doing the same thing to me?

I sat there, frozen in stony fear, much too terrified to cry. Unlikely as it seemed at the time, the day got worse, not better. Russell's screams dwindled to a continuous whimper, the kind of thing you might hear coming from death row during the last meal. Others followed suit. But most of us just sat in our chairs like unsmiling Pillsbury doughboys. One or two misbehaved marginally, but not nearly enough to make me feel more comfortable (especially after watching Mrs. Steprow whittle them down to size).

Thus did I stumble into my scholastic career in Northfield, Ohio. It would continue to be a rather severe struggle for me until well into my fifth grade year. Mom tells me that during the first year or so I used to cry each morning before the school bus came. I don't remember that, but I do remember the anxiety that was my companion much of the time. To this day, I can feel some of the residual dread that automatically begins seeping into my blood when the late summer sighs in weariness of her burden. Mom probably never suspected the shot of icy adrenalin that accompanied her annual first-day-of-school rendition of:

> "School days, school days,
> Golden, golden rule days.
> Readin' and writin' and 'rithmatic,
> Taught to the tune of a hickory stick."

That hickory stick stayed stuck up my spine for many Septembers.

There is a mystique about school buildings unmatched by any other kind of structure in our society. Some churches come close, but not many, and not often. Any one of thirty million of us can walk into our old grade school classrooms, that we have not visited for decades, and be at a total loss for words to describe our feelings. There is no small piece of us here. The eight-year-old within you, which (to your continuing surprise) never quite went away, invested a thousand enormously impressionable hours inside these walls. Standing there a dozen old memory friends come calling: the smell of pine cleaner and crayons, the sight of the cursive alphabet circling the room atop the blackboards, and the sound of tinny bells and a hundred scuffling feet. You look at the room and wonder, "Which part of me left, and which part of me stayed?"

The older the school is, the better its mystique. Mrs. Steprow's kindergarten class met in the basement level of the Northfield Elementary

School in the heart of Northfield Center. With the exception of our well-lit classroom the basement held a *Phantom of the Opera* aura of murkiness, with shadowy stairs giving way to a dark open area that in turn led to the restrooms. Everything was high ceilinged, giving a further sense of being deep in the bowels of a cavernous dungeon. Here the school janitor would hang himself a year or two later, an act that I always thought fitted the mood of the place.

Upstairs, where I would spend my fourth and fifth grade years, the rooms were large and bright, compliments of the vast window surface needed to climb to the level of the high ceilings. In the fall, those windows wore pumpkins and witches, while Christmas saw them dressed up with triangular green trees, silver stars and snowy cotton. The walls hosted the obligatory pictures of presidents, a bulletin board with blue, red, and gold stars by our names, and hooks stolidly holding our crookedly flung coats. The desks still had holes in the upper right corners for the long-gone ink bottles of earlier generations, and were permanently bolted to the floor in testimony to the last vestiges of a time when orderliness came before creativity, when rows ruled rather than clusters. Outside on the grounds we planted a tree each Arbor Day (we grumbled inwardly at not getting the day off), and gambled thousands of puries and catseyes in a two-year long game of marbles ("Last—pegs, shoots, and rounds only!").

The only other old school I attended was the junior high school in Macedonia, formerly the Macedonia High School before the four communities (Macedonia, Sagamore Hills, Northfield Center, and Northfield Village) consolidated school systems. It, too, was a fascinating though unimposing structure, with dark halls and trophy cases containing pictures of dorky-looking World War II-era students. The old elementary school is now the office headquarters for the board of education, and the junior high school was turned into some kind of tiny office park. Sacrilege! Its few little businesses looked as natural there as cats in a swimming pool. As soon as you stick your head into the building your ears hear a thousand faint whispers speaking of dances and homework assignments and basketball games. That building could never be anything other than a school.

First grade was another jolt to my little system. After finally warming up to the snacks and naps and sandboxes of my half-day kindergarten world, I learned that my first grade sentence would be served at Lee Eaton School far up in Northfield Village, which is as far as you can get from my native Sagamore Hills without leaving the school district. No doubt the opening of Lee Eaton was a great moment for School Board members, community planners, teachers and, of course, old Cyrus Eaton, himself, the multimillionaire industrialist who, I surmise, planted a pile of money in this building that would bear his daughter's name. (Eaton, who gained Cold War notoriety because of his chummy relations with Nikita Khrushchev, was one of two reasons people may have heard of Northfield, the other being the racetrack.) I was somewhat less than enthralled with the grand opening.

But the bulldozers that carved up the vacant field in Northfield Village in preparation for the Lee Eaton Primary School were also cutting the beginnings of a road that the community-pride zealots had probably not foreseen. It was a road that led away from the Old Northfield of Friday night football games, Memorial Day parades, and bobby-soxed teenagers at the Varsity (ice cream) Isle, and toward the mist-shrouded outlines of shopping malls, housing developments, and rows of mailboxes bearing Polish surnames. Thousands of other American communities were beginning to cast curious glances at similar roads running through their town squares, wondering at the clamor that seemed to be growing along the far end of that road just over the horizon line. The whole process was tied to the unique terms and phrases of the times, "white flight," "mobile workforce," and "urban decay." In our rural neighborhood, these things meant that dust-covered Glencrest Drive would now be tarred and extended all the way back around to Boyden Road, that the car coming up our road probably wasn't coming to our house, and that kids we played with named Smith and Bean and Mansy would now be joined by others named Kenesky, Subotnik and Papara. The little barbershops, bakeries, and hardware stores that lined "the center" of 150-year-old Northfield began to lose their tidy sense of centricity as the psychological foci of place slipped out into the gray somewhere of a dozen new subdivisions and the new Northfield Village Shopping Center. (Across from the Shopping Center on Rt. 8 I noticed a new restaurant one day, with yellow arches cradling a marquee that bragged that the newly formed chain had sold over three million hamburgers nationwide.)

My class was always at the cutting edge of the turbulence in our burgeoning and over-matched school system. We were shuttled around to seven different school locations in twelve years, including a Baptist church, a Masonic Temple (a large hall curtained into quadrants, a situation that led one mother of a fifth grader to complain that her child was learning more about sixth grade than fifth, thanks to the loud-voiced teacher on the other side of the curtain), and two new school buildings that we helped to christen. The first of those, Lee Eaton, was not quite ready for us in the fall of 1954, a circumstance that stretched my summer vacation for several days, perhaps a couple of weeks. Naturally, I was pleased with the prospect of a reprieve, but the delay really wasn't enjoyable. Lloyd was already back in school in Northfield Elementary (I don't remember why Lee Eaton didn't take third graders that first year), as were most of the other kids. The hiatus only prolonged the sense of doom dripping from the lingering heat and humidity as I sat listlessly inside the house, now bereft of playmates. It was like a temporary halt at the front of the school vaccination shot line while the weary doctor and nurse pause behind a tray loaded with vicious syringes to review some glitch in the paperwork: "Just a moment, young man, we'll be right with you." Right.

The smell of a new school building in the 1950s was as intimidating as the sight of it. That smell would reappear each first day during the succeeding falls, but never again would it cut so deeply into my sensory world. Every part of it was fresh and razor-sharp—the clinging pages of new textbooks, the wax on the floors, the hint of drywall lime, the paint on ten thousand square feet of surface. Then, when the institutional smell seemed no longer tolerable, some bleach-faced kid would make the first human contribution to the essence by throwing up all over his desk. Enter the brow-furrowed janitor with his rolling gray pail and powdery, green deodorizer, and we were at last on the way to making this building into something that smelled like a school.

"Jeffrey, you're in Mrs. Flea's class," said the woman behind the table near the front door of the building. She used the clipped, clear tone reserved for kids who can easily misunderstand the simplest of instructions. "Go down the hall to that classroom just beyond the big clock."

I turned and moved mechanically toward the room that had been so unmistakably identified for me, dreadfully aware that I was exercising my last option as a free boy. Once I entered that room "just beyond the big clock" I would be in for a full year of full days. There would be no going home for long afternoons with Mom this time. This was first grade in the Big House, full term. I was a lifer. The clock was half as big as I was. I jumped as it ticked off a minute by clicking its big hand backward a half notch then forward to its new position, like a marionette soldier presenting arms. How I already missed the warm hum of the electric clock in our kitchen at home. Every minute of every school day for years to come would be announced by the loud, backward-and-forward jerks of that clock and others like it, and they would provide a fitting reflection of my disjointed progress through the halls of public education. "Just beyond that clock" seemed a place that would prove almost beyond my endurance.

Mrs. Flea was an entirely forgettable teacher. For the most part, I have obliged that quality in her, remembering only snippets of first grade, save that memorable long walk down the short hall on the first day. I don't remember her smiling, even one time that year. With a name like hers I suppose that you end up with either a great sense of humor or none at all; I guess she opted for the latter. I was enchanted by something called the "cloakroom," a hall-like sliver of space running parallel to the main classroom with openings at both ends. Naturally, Mrs. Flea expected us to hang up our coats therein (she insisted on calling them "wraps," or sometimes "cloaks"), but we saw this as the least exciting possibility for this luxury that had not been afforded us in kindergarten. Clearly the space would serve much better as the backstretch for chases, a haven for words not meant for Mrs. Flea's ears, and a vantage point for eavesdropping or simply hiding out.

My worst moment came a short time later during that rich piece of first-day ceremony known as "the rules." Mrs. Flea had decided to talk about lunches, and even without the benefit of a G.E.D., I was smart enough to figure out that my sack lunch had marked me as an odd man out that day. A desperate scan of the other kids' desktops confirmed the dreaded realization that most of them were planning to buy what she kept referring to as "hot lunches" in the cafeteria. Mom had packed my lunch neatly into a little white bag. Though Mrs. Flea never came right out and said so, I got the distinct impression that she believed sack lunches were far inferior to "hot lunches." You could tell that by the way she kept saying "hot lunches," as if they were a standard part of American virtue, and the only alternative to them was cold sewer carp.

I might have been able to weather that traumatic lecture had it stopped there. At lunch time, I could have quietly sought out the other first-grade dregs who had brought sack lunches, and we could have collected ourselves unobtrusively at the end of some unwanted cafeteria table, there to nibble away at our shameful fare (trying desperately not to rustle the offending bags), while stealing glances at the normal kids eating their hot lunches. But I was not afforded even that crumb of comfort on this comfortless day. To aid her in making her point about what was and was not preferable in sack lunches (I noticed she made no such analysis of hot lunches), Mrs. Flea took *my* lunch and exhibited its contents before the class. She held up a couple of cookies Mom had wrapped in waxpaper and told the class that cookies were probably not a very good idea; why, I don't remember. She did not say this in a demeaning way, but it made little difference. I don't remember if she analyzed anyone else's lunch or if she thanked me for providing the means for her demonstration. My head was buzzing, and my stomach felt as if it were housing a waterlogged softball. Eating any kind of lunch that day had become a moot point for me.

My luck with lunches did not improve much during the ensuing days. Mom had made the considerable gesture of buying me a glass-lined thermos, a scientific advance I found quite exhilarating since it freed me from my only other dietary dependence on the school, that being those three-cent, half-pint bottles of milk that looked as if they belonged in somebody's dollhouse. Unfortunately, I was not sufficiently impressed with the fact that this liberty was dependent upon my ability to keep the thermos from harm. One day it died where it had lived, in my lunchbox. Of course, it wasn't really my fault. Every grade schooler in the 1950s knew that lunchboxes were only secondarily employed to carry lunches, and that their main value came in banging other kids over the head, or in falling noisily to the floor during spelling tests. You would think that lunchbox designers, knowing this, would have made them with exterior foam rubber cushions or miniature airbags that inflated on impact. But in that age when Ralph Nader was only a few grades ahead of me in school, the lunchboxes were made of thin metal that inevitably assured the destruction of any fragile contents. (Two generations later I would note thermoses had graduated to plastic.)

After my lunchbox's long fall to the floor, I quickly sorted through its jumbled contents and grabbed the thermos and unscrewed the twist-on cup-cap with an instinctive sense of dread. Under the cup, I was confronted by a mound of splintered glass, billowing up like the globe of an ice cream cone. The advanced maturity afforded by my first grade perspective allowed me a full three or four seconds before exploding into tears. It was more than having ruined a wonderful thing. Somewhere in the emotional logic of six-year-olds I managed to draw the conclusion that when I broke the thermos, I had also broken a part of Mom.

It seemed that we frequently had substitutes that year. I suppose that poor Mrs. Flea was often sick with whatever it was that never allowed her to smile. While I have no research to substantiate it, I have a theory that subs in those days were a different and more dangerous lot than today's breed. Modern subs are often entirely competent educators who either choose not to teach full-time or actually prefer the challenge of being plunged into a different environment each time they enter a classroom. Not so, I think, with the subs of 1954—or at least a few of them. These bore all of the marks of the local school system's bottom feeders. Even as single-digit graders you got the impression that many of them saw their occasional calls to service as opportunities to counterattack against the system that (so rightly) kept them out of the full-time classroom. I suspect that the only thing that convinced principals to make those disagreeable early morning calls for subs was the slightly worse prospect of having to sit for the teacherless classes themselves.

I remember one such sub. She marched around under a pile of straw-spiked, black hair sticking out wildly in several directions (a full half-century before the style became faddish) and squinted out at the world from behind stern, dark-rimmed glasses. Physically, she was built like half of a Friday night tag-team match. Somehow she managed to make the absent Mrs. Flea suddenly seem warm and loving. The woman ("woman" was my best guess; I was never really sure) once nearly yanked my hair out when I dared to get too close to the class dollhouse. I felt my feet actually leave the ground. Another of our subs came to approximate a human being only when she was having us sing through a silly song about a fat lady sitting on someone's hat. She would laugh her way through that song, smiling condescendingly at us as if having done us some great favor in allowing us to learn it. "Christopher of Columbus, what do you think of that?"

Macey Johnson was as lively as Mrs. Flea was lifeless. She was not one to let any of her students stand timidly by the edge of the pool. So, in second grade, during Lee Eaton's second year, I got wet. In many ways, the dousing was less than enjoyable, but there is little doubt that Mrs. Johnson helped give me a tougher hide for the demanding twelve-year journey through public education. She had a pretty tough hide herself, and more often than not wore that junkyard dog look that effectively precluded hours of classroom trouble from kids who will always take what you give them. I have noted that it is also a look perfected by harried cops and major league umpires, though it seems to have fallen out of favor with teachers.

I was with Mrs. Johnson when I first met Jerry McCauley. I had asked her if I could get a drink of water and, for some reason, she had put her hand on my shoulder and walked me to the door (a courtesy she seldom extended). Just as we stepped through the doorway together, a blinding flash cut across my bow. It hit my forward striding leg and flew through the air, bouncing to a prostrated stop a good fifteen feet down the hall. The thing was Jerry McCauley. Through the years, I would find that Jerry's horrendous crash was typical of the way he went through life, but at the moment I gave quarter to the suspicion that he might be dead. I had never before seen a human being (assuming Jerry qualified) take such a long dive and land in anything other than water. Mrs. Johnson was unmoved by the spectacle. She reached down and pulled him to his feet, all the while staring him down with that hard look. My first impression was that the boy had lost half of his teeth in the fall, but then I realized that Jerry's teeth were covered by a full quarter of an orange peel. If Mrs. Johnson was concerned about Jerry's health she didn't show it.

"Y'all tryin' to cripple mah boy?" she drawled, jerking on his elbow hard enough to give Jerry the appearance of a puppet badly out of control.

It never occurred to me that my leg might have been hurt in the slapstick encounter. When you were around Mrs. Johnson the only pain you thought about was that which *she* might be administering at the moment. But even then, you preferred any pain to that look. Pain was the end of the line, but that look went on forever.

After primary school, I did not see her for nearly ten years. Then one day, as a high school senior waiting for my bus, I was half spun around by a firm jerk on my arm. The eyes were vaguely familiar, but beyond that I was hopelessly lost. Clearly, though, she had no similar difficulty recognizing me. She stuck her face close to mine, now yielding a good half a foot of height in the process.

"Do you remember the time you said you'd ruther not learn to talk at all, than to learn to talk Southern?" she asked. A trace of the look was still there, but now was greatly softened by a teasing smile. I couldn't believe that I had ever said that to her; I think I must have said it to Mom and it got passed along. I wouldn't have been standing there in reasonably good health in 1966 had I said anything like that to Mrs. Johnson in 1956. I also couldn't believe that she recognized me. It was, after all, a fair bet that I had changed considerably more than she had during our separation.

I guess one of the marks of good teachers is that they remember their students. Thirty-four years after leaving third grade, I sheepishly knocked on the back door of Mrs. Oviatt who had ushered me through that last year of third grade at Lee Eaton. She opened the door and instantly greeted me warmly, even though I had not given her any warning about my visit. Before I could get halfway through my prepared remarks intended to reassure her that I wasn't someone trying to sell her something she waved me off with the comment, "You still have that same smile you did as a little boy." Mrs. Oviatt was a high-class teacher, the only one my mother ever specifically requested for me before the start of a school term. Lloyd, and I think Jack, had her before me, which made her the only grade school teacher to have had all three of us. It was fitting that this human monument to our early educations was also connected to some of the earliest and most prestigious settlers in our neck of the Western Reserve. Her husband's name may be distantly connected to William Oviatt, son-in-law of Jonathan Hale, although there were other Oviatts in the valley at an early date. (Mrs. Oviatt downplays the probability with the comment that those Hale-Oviatts of William's time were the ones with the money.) Just around the corner from her house sits the small building that once was Little York School, one of the first in the area.

I did not get back to Lee Eaton School for fifty years after I left third grade. It is difficult to accept that it is, as schools go, probably past mid-life. I can picture it as nothing less than a new building, still buzzing with sounds of our third-grade operetta, "The Honey Pirates," (I was Drowsy Drone), and softly smiling at my first case of puppy love (Jeri Lu Davis, where are you?). For a long while I thought best that I never go back there. That way, no one could ever tear it down.

~~~

There were other buildings in the valley. These, too, housed people, but only incidentally, and only to the extent that they were necessary to do the business of business. Arguably, the Western Reserve was the heartbeat of America's industrial development. At Cleveland the fabulous iron ore wealth of the Upper Great Lakes met up with the rich streams of coal flowing from the Appalachian hills in Ohio and West Virginia to create the steel that would be the backbone of the nation's industry for a century. Production of the automobile, which remade the U.S. economy in the twentieth century, was originally identified with Cleveland more so than Detroit. In my youth, the canyon-sized Ford and Chrysler stamping plants, the latter covering some two and a half million square feet, were to be found within a few minutes' drive of our rural Sagamore Hills homestead. And, of course, Cleveland's boilerplate industrialist, John D. Rockefeller who, when asked how much money was enough, said a little bit more than what he had, set the United States on the road to the mixed blessings of the petro-economy, birthing issues still hot to the touch. To the south, in Akron, men like Goodrich and Firestone built structures to house the rubber center of the planet, while Ferdinand Shumacher built and, when disastrous fires destroyed them, rebuilt his mills that fed generations of Americans their morning portions of Quaker Oats.

Even our rural nook of the valley had something of an industrial heritage. A few short years after Northfield's founding father, Isaac Bacon, settled into a cabin on the northwest side of the township, George Wallace was coolly appraising the power producing potential of magnificent Brandywine Falls to the south. A sawmill went up in 1814, followed by a grist mill, distillery, and a woolen factory. In 1816, Brandywine (locals argue whether the name derived from the Revolutionary War battle or the area's most successful early product) was commonly assumed to have a brighter urban future than Cleveland. The foundation and some of the stonework from the mill still sit on the south side of the falls, carefully tucked in among the walkways that now line the canyon walls (again, compliments of the federal park people's determination to amplify the valley's rich heritage), reminding the world that capitalism arrived on the heels of agriculture here.

By all logic, there should be little of sentimental value in the huge boxes that housed the best and worst of the nation's commercial might. They were rudely imposed on the land at a time when no one gave a thought to environmental implications. In Akron, the old German's Quaker Oats mill has been incorporated into a classy downtown shopping mall, and some of the huge rubber companies sit quietly in their abandoned kingdoms. In their heydays, these monsters fed tens of thousands of people, but such was not their goal. Their goal was to see the making of huge fortunes, which, in turn, would fund more such buildings. When it came their time to be vacated, or renovated, or excavated, why should anyone care? But, strangely, at such times we do care. Even the most dehumanizing of these structures was filled with human stories. As someone once said, God created men and women because he loves a good story.

Solon Foundry was my place for such stories in the mid-1960s. It was also, I should note, a paycheck for five summers during my undergraduate years, but now that seems the least important part of it. Solon Foundry was not a structure of distinction. No one was likely to confuse it with the graciously landscaped Western Electric plant one minute down Rt. 43, nor did the city care to profile it on the front of chamber of commerce literature. The building—I take some liberties in using that word—was an erector-set patchwork of brick, glass, wood, and iron strung together along the railroad tracks behind a raggedy shopping strip and Roger's Bar.

No building in the history of architecture was devoted more to pragmatics and less to aesthetics. It went where it had to go in order to produce aluminum castings, and no further. Most of the glass windowing back in permold was busted out, but no one seemed concerned about replacing it. After all, the hot furnaces heated the great room in winter and it needed venting in the steamy summers.

The foundry was one of the rare forms of American business that came to us as an aged art and changed very little afterwards. In many ways, foundries were as primitive in 1969 as they had been a century earlier—or four thousand years earlier, for that matter. The basic idea was still to melt metal bars in open vat furnaces, then pour the molten contents into molds. In our case, the molds were often interlaced with sand cores in order to produce detailed castings for a variety of uses. At Solon, these uses included parts for ladders, sidewinder missiles, and engine housings for military helicopters, among many others. A series of tasks accompanied the making of the final casting—knocking out extraneous sand and metal, x-ray and zyglo processing to discover faults, tolerance testing, various machining activities—but all of this was only window dressing on a process used by the Babylonians and Egyptians long before Christ, and much of it is there in Georgius Agricola's *De re metallica* published almost five hundred years ago. Once, one of the office workers at Solon showed me some pictures of a new experimental foundry boasting a great array of mechanized movement the focus of which was a series of robotized arms, each with a ladle for a hand, dipping molten aluminum out of clean furnace pots and precisely pouring it into neatly spaced molds. There was not a human being in sight. A short time later the place went belly-up. I guess foundries were meant to stay primitive.

The Solon Foundry of my late-teen work world was the second place of that name I came to know in my youth. The first was the one I saw through the impressionable eyes of a little boy whose dad happened to be the foundry's production manager. That Solon Foundry was a wondrous place for Lloyd and me. The singing of the band saw and the blankets of heat from those incredibly dangerous furnaces pressed deep images into the soft clay of our minds. We figured Dad must be pretty important to be at the center of such a big, rough world. If Mom had need of the car during the day, we could look forward to riding with her to pick up Dad at the end of the day, an event filled with little traditions that could only have meaning for wide-eyed kids. Dad would let us walk with him as he made his last pass through the "shop" for the day, checking the progress of jobs that might translate into a rude call from a customer on the morrow. Lloyd and I were always fascinated by the speed of Dad's strides as he made that last round. I suppose the drive came from the "production" part of the production manager in him, but at the time it was just a part of a game in which we sometimes had to break into a trot to keep up. (Forty years later, I would find myself walking rapidly through the calm, carpeted halls of my research workplace, even though time was seldom of the essence.) Occasionally we had the rare treat of riding with him as he backed the company truck into the single loading dock for the night.

These exotic pleasures were, however, secondary to watching the foundry softball team. Those productions were as aesthetically pleasing as the foundry was pragmatically productive. The ancient human drive for pageantry exploded out of the smoky foundry confines and onto the beautifully manicured green ball field in flashes of color, speed and strength. I can vividly recall most of the brilliant uniform colors: the menacing blues and whites of arch-rival Cyril Bath, the soft yellows of always-beatable Colorcraft, and the razor sharp blacks-and-whites of Solon. The game was fast-pitch softball, and it moved so quickly that I often had trouble sorting out the happenings. Once, during a tournament of some kind, they had to play under the lights on a far-away field. Having never seen any night game other than ones involving the Indians, I feverishly asked Mom if the game was being televised.

One living symbol of Solon Foundry was not human but rather feline. The shop cat was a foundry fixture from my earliest memories. It was not, of course, the same cat throughout my fifteen-year association with the place, but it might have been, had appearances counted for everything. She was usually a gray calico, a good Darwinian choice amidst the perpetual gray of the aluminum stains and smoking furnaces. Lloyd and I used to wonder if this was a safe enough place for a nice kitty, but Dad just shrugged and hinted that she was every bit as tough as the men who shared the filthy floors with her. There was an unspoken respect between the men and that cat, each realizing that they were all here to do tough jobs as best they could. Occasionally, one of the men would wordlessly toss a scrap from his lunch bag. She usually accepted this gift if it fell within the confines of her quiet dignity—things could get tough for a shop cat—but she was no beggar, and was more likely to be seen in a business-like trot to some obscure corner with a mouse dangling from her hard mouth. No one went out of his way to look out for her; that was understood. Once I saw her seemingly trapped in the loading dock as a truck was backing in. Buzzie was at the top of the dock six feet above her, guiding the truck driver back in, occasionally casting a disinterested glance down at the cat, which seemed determined to get the last whiff of meat out of some deserted chicken bone. Buzzie never changed the pace of his job, nor did she change hers. The truck kept coming. When it was about four feet from her, she calmly gathered her sleek body into a spring and bounded up to Buzzie's feet with an incredibly easy jump. They were both entirely matter-of-fact about it all, even afterward. Buzzie had a truck to unload, and she had her face to clean. But the endings weren't always happy ones. One day, I saw one of her kittens dragging the back half of its body behind it. It had not been quick enough to escape a falling drum or a careless footstep. The hard job of drowning the kitten fell to Dad, though I don't know why.

The hardness of the foundry was a given. It drew to it men with a peculiar, steely look in their eyes, bearing names such as Joe Svanda, Mike Grubich and Bill Cross. With the exception of this "foundry look," these men did not resemble each other, but because of it I always thought of them as being somehow related. There were other looks, of course, the soft, sad looks of defeated men, the bleary-eyed looks of the alcoholics, and the happy looks of those eternal saintly types who can find a piece of God everywhere. But it is that piercing foundry look that has stayed with me over the years. In it I saw the ultimate toughness and defiance of men who, for a while, at least, were determined to show this filthy sweatshop that they were up for its tests.

It is still inconceivable to me that someone could look upon a foundry job—or at least some jobs in the foundry—as an indefinite commitment, something you would leave home to do each morning in the same way you take out the trash on Thursday or pay your taxes in April. For those of us working summers it was just an interesting and laborious excursion into the world of work. For $2.37 an hour we were treated to some colorful people, the chance to play some good softball, and a situation we could and would walk away from come September. It was an annual, finite challenge that we, in the cool of our dorm rooms, could look back upon and savor precisely because of its finiteness. Survival tests are shorn of heroics when their length is a predetermined thing. Had there been no back door for me at Solon Foundry, waiting slightly ajar in anticipation of my predictable escape, I doubt that I would be writing quite so affectionately about it now.

Even under these controlled circumstances, I was occasionally nicked by the hard edges of the place. One summer when Solon fell into a crack of the Viet Nam-inflated economy and jobs became scarce, Lloyd and I had to work the dreaded "shakeout" on the second shift. A half hour after a particularly large casting had been poured into a sand mold it had to be hoisted up and freed of its scalding sand by beating on it with wooden mallets. The task itself was not difficult, but the resulting environment was. When the steaming sand belched onto the floor you found yourself in a sea of black smoke and haze—no real problem unless you wanted to see or breathe. My first night on that job I became dehydrated and learned what people mean when they say they "couldn't stop drinking water." It was that same summer that I tried my hand as an oven tender. The small sand cores were loaded onto a large steel-poled grill, probably six feet by eight feet and containing several shelves, which then had to be jacked up on a wheeled platform and put into the ovens. Since each of the ovens could house two of the grills (end to end), and since they were never turned off during the shift, taking the first grill to the back of the oven meant that for the agonizing moment it took for the pneumatic jack to lower its load to the floor, the oven tender baked right along with the cores. My first trip in, or, more accurately, my streaking retreat back out, must have looked like a scene out of a Chaplin movie. I learned after that to wipe the sweat off of my face before entering (it had a way of boiling on your face) and to keep my face protected by my arms while the jack was doing its slow work. But these were just the comic relief scenes in the foundry, something to give the veterans a good belly laugh. During my last summer at Solon, part of my duties required me to take injured workers to the doctor. Molten aluminum is not very forgiving and never draws a laugh. Beneath its splash all tortured skin looks the same, black or white.

The foundry men formed an interesting demographic hodgepodge. There were some WASPs, of course, and a few foreign-born workers, often from Eastern Europe. There were a greater number of blacks in Solon's labor force, and a similarly large group bearing hearts so passionately yearning for West Virginia and Kentucky that they could barely wait for the 3:12 p.m. Friday whistle to free them for their weekly dash down the newly completed segments of I-77. The handles for these groups wore some of the rough edges of the times and place. Some of the crude monikers were "DPs" (for "displaced persons") "Billies" (without the "Hill...") and "Coloreds" (or worse). As for the summer help, we were just known as the kids, or, when one of us made some stupid mistake, a *college* kid. In face-to-face communications, nicknames were often favored, but so were some of the formal names. When riled, the guys would resort to any noun that would bear the weight of a standard list of adjective obscenities.

The men sometimes wore odd names: Lonus Justice, Iree Lardell, Pleze Martin, and Hazel Brown. All of us shared a pretty low rung on the social ladder, though everyone knew that the college kids were on the way up. Surprisingly, the mix worked rather well. Amazingly, I don't remember one fist fight during those irritatingly hot summers, and the arguments always had more to do with the job at hand than the skin color of the hand. At lunch, the men did divide into their predictable groups, but these were no beds of isolated tension. During my last year of work at Solon, when I was the only college kid there, I used to eat with the black guys. This was no great social statement on my part. I just happened to feel more comfortable with several of them I had known for five summers, especially those on the softball team. Besides, they had a lock on the only picnic table. When a group of locals sped by one noon hour and shouted their classy one word assessment of our gathering—"Niggers!"—I took a good bit of teasing from the guys for having failed to disqualify us from the insult. Suffice it to say, I learned a good deal at Solon Foundry about that universal predicament of human kind.

The foundry was a remarkably sensual world. Virtually every value judgment was based on some aspect of seeing, hearing, touching, smelling, or tasting. The constant preoccupation with sex rivaled that of the high school locker room—not as loud, but more crude. Hence, the language, the porn paste ups, the endless jokes; maybe the guys needed that cheap distraction in that place. But there is nothing quite so sad as a middle-aged man trying to suck some faint taste of pleasure from foul jokes and lewd allusions that have long since dried to dust in his mouth. His laughter is forced, desperate, a taunting reminder that his younger years could have been invested rather than squandered. Now he has little to show for nature's most universal challenge except debts.

"How old are you, Little Larry?" one of the middle-aged Poles asked me, assuming that my dad's name was good enough for me, Lloyd and Dad.

"Nineteen," I said

"That's a damn good age to be, nineteen," he said, having no success at covering the pain in his eyes. Nineteen is the peak of life when sensuality is trump.

But the foundry itself was not a natural encourager of humanity. Its very filth and stench provided a perpetual sensory reminder that this was life in a low rent district. It was, of course, a job, but most people found work during the Viet Nam era. This was nobody's career dream, just a living. Small wonder the men took what little sensory enjoyment when and where they could.

The foundry's smell is one of the most powerful images I retain of it. For years Dad brought it home with him every night, even though he worked in an office away from the great furnace areas where the smell would be steamed into your pores before noon. Like Upton Sinclair's Lithuanian immigrant, Jurgis, (*The Jungle*) who found that the foul work of the fertilizer plant stayed with him for six months after he quit, Solon Foundry could lay a pretty deep mark on you. (I always wondered what would happen if the detergent advertisers tried out their stuff on one of my permanently grimed, foundry shirts.) Once, years later, I hopped onto a city bus in Columbus and was immediately assaulted by the unmistakable smell of foundry. It wasn't hard to find the source. He was sitting about halfway back on the right side, with empty seats all around him. But that smell tripped a flood of good memories in me. Each night after that as I boarded the bus I eagerly hoped he would be there. It was my instant transport back to a time, a place, and a people I had loved in a unique way.

If blatant sensuality was the limitation of these people, spontaneity and openness were compensating virtues. There was a rough honesty about the men that I have not found elsewhere, even in churches and schools. They had apparently realized long ago that they weren't very good at hiding things or fooling people, so they made no effort to do so. What you saw was what you got. They wore their passions and their senses of humor on their sleeves. Odd that so much laughter would flow from men who seemingly had so little reason for it. But it was there every day, loud, booming swells of it filling huge rooms already saturated with industrial noises. And endless teasing, jawing, and little games going on over the tops of work benches and up on the bulletin boards. Sometimes the place seemed more like Santa's workshop.

Then there would be an explosion of obscenities and shouting. Men were into each other's faces as if someone's family was being threatened. There would be a great deal of gesturing and posturing to accompany the verbal violence, finger stabbing, and swift kicks into empty air or at an unoffending barrel. Then, just as suddenly, it would be over. Someone, with a great deal of effort, might prolong his sullenness for an hour or two, but the big-voiced laughter was usually back long before then, bullying the bad feelings back into dark corners. Funny, but a few loud words spiced with an obscenity or two injected into my present, soft-cornered world of niceties is enough to raise instant moisture on my skin. Back then, it might not have even drawn my eyes from my work. The difference, I think, is that there was no unknown element among the foundry men. Every inch of their character had been fully exposed a thousand times over. There was simply no material for long weeks of intrigue and whispered gossip and quietly nursed wounds. Two men might hate each other, but if they did they quickly got around to saying so, after which they could go about their business.

But enmity was not the norm in the shop; an odd sort of camaraderie was. It was a mutual reliance born of necessity, not luxury, as if the men knew that life on the ragged edge left little room for false pride. Secrets their wives never suspected (maybe just as well they didn't) were openly passed back and forth among guys who weren't ashamed to be bewildered about life. Louie, a fellow ballplayer who had once confided to me that he had to leave Alabama because he had gotten into a little trouble (he had killed a guy), told me that coming to work was his one hope of beating some severe daily depression. (He didn't say it quite that way.) Simply put, his spirits rose once he began the ritual of insults, pranks, raw jokes, and arguments that comprised the foundry life. I think it was that way for most of them.

There was one guy who was a paradox, who fit into the foundry both beautifully and not at all. We all called Phil Scott "Red" for the usual reason. (Nicknames at the shop weren't particularly imaginative—Slim, Baldy, Shorty.) Redheads were in short supply there, but Red would have stood out anyway. He was the one person at Solon Foundry who was a crossover, who easily walked the disparate lanes of the foundry worker and the college kid. Unlike Lloyd, Stevie Grubich, and me, Red did not turn in a final time card each September and head off for some distant school. He was pretty much full-time at Solon, working for Dad, as a matter of fact. But he was also wrestling his way through night courses at Cleveland State University (I think) in the manner of lower-middle class kids from large Catholic families.

Before Red went to work for Dad, I could probably count on one hand the number of times Dad brought home to us any human interaction that had blessed his day at the Foundry. His work as production manager was enormously stressful, and a part of Dad had long before accepted the premise that this was not the arena in which his human needs would be met. But that changed with the arrival of the Great Scott. Dad took much delight in passing along to us Red's antics from the day, or things that he said, or ever-unfolding developments in the soap opera scenarios from Red's love life. "That guy's always got something going," Dad would say, shaking his head, but with eyes alive with approving interest. To be sure, part of Dad's feelings for Red was probably identification with his own virtually grown sons. Lloyd and I were natural friends with Red during the summers, even though the color of our collars separated us during much of the work day, but during the short winter days, when the Foundry's furnaces turned from enemies to friends, Red must have been a warm reminder to Dad of his two boys far away in Tennessee. And, no doubt part of it was Dad's admiration for the qualities in Red he would have respected in anyone: intelligence, healthy ambition, a willingness to work hard, and an indefatigable sense of humor.

But mostly, Dad loved Red for the same reason we all did, because of his Alice-in-Wonderland zest for life. The guy acted as if we are actually supposed to enjoy our pass through this thing called existence. Sometimes, I strongly suspect that the first words we will hear from the Lord upon our arrival at eternity's threshold will *not* be "Well done, thou good and faithful servant," or "When I was naked, did you clothe me?" but rather, "Why didn't you spend more time in joy? Why were you always finding some excuse to quit or hide or sulk, or conversely investing thousands of hours in endlessly repeated and twisted activities that were only meant to momentarily amuse you? There was so much there I wanted you to *enjoy*."

This was never Red's problem. He spent so much time in the world of joy that it tended to follow him around like a bright glow. Any time he approached a group of workers two things happened: first, the men eased off of their work, laying down tools or at least slackening their pace; and second, they smiled. I don't think there was one corner of the shop where Red was not welcome (well, maybe where men were doing piece-work). I don't think there was a type of personality that he could not, sooner or later, reach with a piece of his perpetually soaring spirit.

We shared much with him. A tree's shade each day at lunch, his morning notes at the time clock ("What, late again?"), the foundry softball team, even the up and down world of his relationship with his girlfriend, Mary Ann. As the social world of our high school days misted away he became our best friend back home. We were all somewhere between 19 and 22, ages during which friendships have few limitations, and people seem to fit snugly into the roles we envision for them. Red was a buddy, Dad's assistant at work, and our link to the world of summertimes at home. A solid structure in our lives.

One seemingly benign Saturday, Dad, who worked Saturday mornings then brought a pile of homework home for Sunday night, drifted quietly into the house as Lloyd and I were washing out a late sleep with some brunch. Until later I didn't recall that he had spent a couple of odd minutes just standing around, as if unsure of where to go in his own house. Then, suddenly, and the only time I ever saw him do this, he was weeping. Before the shock of his crying could fully hit us we were blasted into numbness by his quavering words: "Red was killed last night in a car accident." The story had been laying six inches away from us in the morning newspaper. I may have even touched it when I routinely checked the Dodgers' box score, but we hadn't known it until Dad told us on a bright Saturday morning in June.

Lunch time at the foundry. Solon, Ohio. Me, Lloyd, Red, and Steve Grubich. Friendship short-circuited. *Photo from author's collection.*

Like any youth seeing death close up for the first time, my orderly world listed a bit when Red died. Some of that was probably inevitable anyway in that painful tunnel of experiential life that leads, as A.E. Housman noted, from one-and-twenty to two-and-twenty. Images blur, roles change, and people no longer fit into the neat cubbyholes young minds create for them. But the shop took Red's death in predictable stride. "I guess youse guys know Red got kilt," Mike Grubich said to us as he handed out black armbands to sew onto our team uniforms for the remainder of that season. That was about it. There were no visible tears, no long orations, no poignant trips down memory lane. The men in the foundry, unlike Lloyd and me, had been down roads like this before. The band saws and the furnaces went on, which meant that life did too. Like the shop cat, Red had been there one day, gone the next, and there was nothing anyone could do about it now.

That may seem harsh, but it was probably not a bad lesson. I know how easy it would be to romanticize Red's life; I have done a bit of that here already. In the less kind realities of long living maybe he and Mary Ann would have divorced. Maybe he would have had a drinking problem. We probably would have lost touch with him, as is the way of things, and settled for rare bits of news about his uneventful life filtering back to us on the whispered wings of friends or Dad's bumping into an old foundry acquaintance at a shopping mall. Maybe he would have quietly despaired of life, as so many do, and settled in for a long stretch with the nightly TV and a six-pack, tuning out the progressive irritants of quarreling children, stacks of windowed mail, and phone messages from his doctor's office. Maybe.

But our world with Red had no maybes. He is still a twenty-year-old redhead with a Howdy Doody face and a perpetual grin, and he has been sealed, along with pieces of my youth, in a slouching, gray foundry building that no longer exists. When I revisited the location a few decades later there was not even a scar on the spot to suggest that here men once sweated out the best hours of the best years of their lives to do one thing: produce aluminum castings. In its place stood a snaking line of small businesses and shops tucked into neat, little rows such as can be found off of the main squares in ten thousand American communities. They sat there remarkably unembarrassed by the fact that there is not an ounce of cohesion in the entire park (a muffler shop and a chiropractor's office blandly stared at each other), and unperturbed by the prospect that tomorrow's change of the sign out front would consign them to instant oblivion. The railroad path still ran along the north border of the property, but carelessly strewn ties and a deteriorating gravel bed betrayed its retirement. I searched in vain for some clue of what happened here a generation before, an ancient piece of aluminum splash embedded in the ground or a work glove lying in a ditch, but there was nothing. Not even a surviving rose from the small beauties that lined the neighbor's fence, and that seemed to me so oddly out of place as they arrayed themselves before the unappreciating eyes of the foundry. No, there is nothing left. Solon Foundry, like the shop cat, and like Red, just didn't show up for work one day. Probably no one shed a tear when the bulldozers pitched into the place. I suppose that is all right; the guys at the foundry would have understood that.

For me there is only the memory of a building locking away a treasure that I dare not trust to bricks and mortar and lumber.

Notes

1. Margaret Manor Butler, *A Pictorial History of the Western Reserve: 1796 to 1860* (Cleveland: The Early Settlers Association of the Western Reserve, 1963), 23-24.

2. Joseph D. Jesensky, *An Archaeological Survey of the Cuyahoga River Valley* (Northampton, Ohio: The Northampton Historical Society, Inc., 1979), 18.

3. John J. Horton, *The Jonathan Hale Farm* (Cleveland: The Western Reserve Historical Society, 1961), 61.

4. Jedick, "Euclid Avenue," 210-211.

5. Some of the castings were made in permanent molds, the others in sand molds.

4
Straight Lines in Curved Space
The Railroads

Water damage to the Cuyahoga Valley Scenic Railroad. Despite the appearances, there are no straight lines on a round earth. We must all bend with the curves—and make some of our own. *Ian Adams.*

The valley's affair with the canal-killing rails • The tracks of a life time—the short lines of Errett W. Knowles

A train horn moans in the night. It is meant to be a warning from a sleek diesel approaching some anonymous intersection, but the sense of urgency never makes it to my ears. I hear only the plaintive cry of a night creature as it plods through its nocturnal duties. The sound of the horn begins slowly, like the wizened groan of an old man rising out of bed, and ends with a fading retreat as the throaty, Doppler dissonance echoes into every seam of the landscape before bouncing gently to a stop in some unseen field. While it lasts, it carries with it a sense of both greeting and passing, like a friend waving from afar, glad to see you, wistfully sad that he cannot stop, unhurriedly but inexorably driven by the demands of his labor.

That train horn in the night is one of the most poignantly peaceful sounds I know. Odd, I suppose, that it should come from a mechanical monster belching a threat. Most people who have bothered to sort through their repertoire of peaceful sounds would probably choose crickets, or rain, or a song. As a child, I heard a great deal from all three of those; they were good and comforting friends. But the train horn holds a special place. It easily cuts through to me in the hotel rooms of Washington or San Francisco, and gently nudges past a dozen worries and a hundred other sounds to join me in my own bedroom in Columbus. Perhaps it is the steadiness that appeals to me. Somewhere out there (you can never tell from what direction the train horn sounds) in the midst of the drug deals, huddled homelessness, and wrecked relationships is a sound that takes a perfectly straight path through all of it. Confident of its own strength and destination, it rolls easily on by, never hurrying, never stopping, never looking back. The blast on the horn is a simple statement of is, not has been or will be. Here is certainty seldom found on the tortured, twisted streets of the city.

But it is also more than that. The horn is a perpetual linkage with my childhood, one of the very few things that seems not to have changed during the cataclysmic changes of my life. The slow blast from the Conrail engine moving steadily through Columbus is indistinguishable from my memory of the B & O coal freighter moving up the Cuyahoga Valley toward Cleveland, or the closer New York Central diesel rattling up the roughly parallel tracks only a thousand feet from our door to the northeast. The sound is an eternal hub that reaches out along the rails of my life tracking me back to my childhood, that central part of me—of all of us—that we fancy we escape, but that we actually orbit. The train horn is the heartbeat of that childhood.

Yet, we don't see our heartbeats, and so it was, by and large, with the railroads. They were assumed rather than defined. We picked berries up by the railroad tracks, and the first time I drove our family car I managed to hang it on the humpty-dumpty wooden bridge looping over the tracks on Boyden Road, but neither of these activities (or any others) made the tracks an indispensable part of our lives. They just happened to be there. Had they not, the berries would still have grown, and I would have found something else to hit. As kids we never stoned the boxcars, or learned to recognize the engineers, or put sticks on the tracks. That lonesome train horn in the night is really a bit ephemeral, almost mythical. There is nothing much behind it, save for its considerable capacity as an emotional tripwire.

~~~

It is just as well that we steered clear of the railroad folklore. While some railroad hobbyists might scream their disagreement, the truth is that the whole venture was a pretty gruesome business in its heyday. Railroads represented the best and the worst of America in the 1800s, and even the best was defined strictly in terms of steel and speed and dollars. It was a century of slashing power and unlimited personal fortunes. Bigger always meant better, as did more, and if some human beings got caught in the cracks and bends of the fast turning machinery, well, that was the price for the better world people unquestioningly assumed was in the making. Two World Wars, a Depression, and an Atomic Age later that assumption would be hauled back under the microscope for a better look, but there was no time for that in 1850. Western Civilization was feverishly feeling its way toward the mist-shrouded peaks of its destiny. Imperialism was not a dirty word; we even twisted Darwinism into a useful, political rationale for what we were doing to people throughout the world. Everyone was expected to see the huge benefits of the exhilarating ride. And our rapid means of transport into the glittering possibilities of the twentieth century was the locomotive. At the time no one thought to build any sentiment into the process. It was just raw power on steel wheels.

There was time for sentiment in later years. Retrospection always encourages selective memory. Remembering the passenger trains running up and down the valley, one account recalls the brass-buttoned conductors, swinging oil lamps, and twenty-five cent fares that framed the portrait of an excursion from Cleveland into the country on pleasant Sundays. The memory included the observation that, "Early passenger cars had red plush seats well covered with cinders and soot from the open, unscreened windows."[1] That was the railroad's footnote on those soft summer trips into the country, those cinders and that soot scarring the plush red seats testifying to the world that while human enjoyment was not deliberately precluded in this business, neither would it stand in the way of profit-making.

The development of the railroads in the Western Reserve was a fascinating mixture of fast and slow. Cleveland, falsely secure in its position as Queen Bee of the Ohio & Erie Canal, was slow to realize—or, perhaps, admit—that her canal coup of the 1820s could so quickly be eclipsed. A railroad charter grudgingly picked up by the city in 1835 lay unused for a dozen years. When work finally began it came in the form of one man with a pick, shovel, and wheelbarrow scraping a path from the foot of the Superior Street hill at the river in the general direction of Columbus and Cincinnati. This was the beginning of the great 3-C Railroad, born into poverty, lethargy, and ridicule. That solitary worker was all the railroad venture could afford at the moment, but he was needed to keep the thing going continuously (save for a rest on Sundays) in order to keep the charter alive.[2]

Ironically, it was Alfred Kelley who finally got the 3-C started on its way. Logic dictated that Kelley should spend his later years emotionally justifying the work of his youth, the Ohio & Erie Canal, as is the way of most men. Comfortably established in Columbus, Kelley remembered his malarial days as the circuit rider who, more than any other man, made the canal happen. Every ache in his aging bones and every lingering chill or fever reminded him of the price he had paid for the great work, just as every blast from a locomotive's whistle told him that the railroads would kill his great work, probably without the decency to wait for his own death. Nevertheless, the desperate Cleveland railroad men turned to him when it seemed certain that the Cleveland-Columbus-Cincinnati Railroad would never be anything more than a piece of paper in somebody's vault. After an evening of resistance, Kelley accepted the job, and the 3-C was on the way to being built. Like the railroad itself, Kelley was no sentimentalist. Someone else would have to nurse the nostalgia for his Ohio & Erie Canal.

To Cleveland's urban rivals around the state Kelley's second transportation success must have been a more bitter pill than the first. For two decades after the canals had bypassed Sandusky, that city had been stoking the fires of rage—publicly during construction, quietly after the completed ditch immediately acquitted itself so brilliantly. Not surprisingly, the folks of Sandusky were quick to pick up on the prospects for prosperity and vengeance contained in railroading. Using funds from a rare windfall granted by Congress on the eve of the financial downturn of 1837, Sandusky eagerly ran for the rails. This effort, too, was marked by the slow and comic quality of the early roads but, as William Ellis noted, it still gave them a leg up on their arch rival to the east: "It may have been a rickety strap-iron, `shake gut' line, plagued, as always, with snakeheads and shifting roadbeds; it may have been horse-drawn and only thirty-three miles long, but, by Godfrey, it was a *railroad*."[3]

The early trains were known as "Satan's Device" to some because of their ungodly speeds of fifteen miles per hour. The Reserve's biographer, Harlan Hatcher, noted that at least one city tried to resist the railroad's reach with the old chestnut that had God intended such outrageous speeds for his people he surely would have mentioned it in the Bible, and concluded that the whole thing was another plot of the Great Deceiver "to lead immortal souls down to Hell."[4] Nearly two centuries ago, that view did not seem so ludicrous. The rolling boilers that passed for early engines terrified children and horses and were, by any measure, a threat to anything within reach, be it a pedestrian or a dry bed of twigs. Hatcher noted that the strap iron attached to the wooden stringers, the first alternative before the production of full iron rails, would sometimes come loose and stab through the floors of passing trains to skewer unsuspecting passengers.

Worse, the railroads had a way of bringing out nonsensical notions in even the most sensible of people. The Ohio Railroad Company, without cracking a smile, set out to build a railroad on stilts through the swampy lands of northwestern Ohio, an escapade that drew serious money from serious men, but that produced nothing more than several miles of artificial stumps. Only a little less ridiculous were the gauge wars among various lines that would see track widths varying from four-and-a-half feet to six feet. This necessarily limited freight hauls and required frequent loading and unloading of cargos. Sometimes the gauge differences were deliberately created. Alfred Kelley nearly lost his life in Erie, Pennsylvania, when he forced a single gauge on that city in his irrepressible efforts to see a singular line driven from the Reserve through to New York. The people in Erie had mandated the differing gauges as an economic guarantee of their city's vested interest in the business of loading and unloading trains. Nor was nonsense the only human frailty encouraged by railroad fever. When Marvin Kent, who lent his name to a neighboring town in the Reserve, desired to build a railroad from the Cuyahoga to Dayton, he named it the Franklin and Warren Railroad, even though neither of those places was anywhere near his secretly planned termini. Had he stated his true intentions, the railroad barons of his day, men who knew a thing or two about corporate raiding, would have gobbled up his little effort. Locals like Kent were not allowed to dream about anything bigger than short lines.

But the railroads were not destined to forever remain slow and silly. Those two horses that pulled stone, lumber, and a few passengers down the couple of miles of the Cleveland and Newburgh Railroad in 1838 had no more permanent part in the railroad story than the trucks that carried the space shuttles to the launch pads have to do with our Solar System exploration. There is a well known picture, probably post-Civil War, that says everything about the railroad's monopoly on speed. It shows a slanting locomotive, its image artificially blurred by its knifing speed, rocketing past a sleepy canal boat plodding along a stretch of slack water. The message from the railroad people was unmistakable. The canal, in fact, was forced into slave labor for the bright career of its competitor and successor. Rails, ties, and even engines were shipped by canal boat to terminus cities where ribbon cutting and spike driving ceremonies awaited the cargoes, not the carriers. I wonder what the old canal captains thought as they surveyed their freight on those peaceful drifts to their own destruction. I wonder if they could sense the grating noises that lay latent in these pieces of iron and steel, noises that would doom canal boats like theirs to rotting graveyards where weeds covered the shallow, dry ditch, causing later generations of exploring children to wonder how a boat had come to be left in the middle of a field.

Nor were the railroaders particularly gentle in turning out the old dowager. Since canals and railroads were equally interested in straight lines and low lands, it is not surprising that they both selected the same routes. The Valley Railroad and the canal bed are often within a stone's throw of each other as they track toward Cleveland, though sometimes the Cuyahoga shoulders its way in between the contestants. At first, where rails and water had to cross, the railroaders constructed swing or draw bridges across the canals, but this expensive courtesy was quickly dropped when political clout shifted decisively in favor of the rail people. This meant that the canal boats had to unhitch their towlines so they could pole under the fixed bridges. But the greater nuisance was the low bridge threat to unsuspecting passengers who would find themselves in the canal at a moment's notice. (There is a century-old vaudeville routine about the unmasking of a would-be socialite who betrays her lower class, canal-boat origins when some wit at a party yells, "Low bridge!" and she dives instinctively to the floor of a fashionable parlor.)

As it turned out, Cleveland's infatuation with the canal did not last much longer than Kelley's. After the Civil War, the city impatiently pushed its aging bride three miles south from her Merwin Street link with the Cuyahoga at Lock 44. Her bedroom, locks 43, 44, and the large canal basin between them, which served as the foci of the city's canal activity, were obsequiously offered up to the lusty, well-railed lines of the Valley Railroad. The railroad would bring to the city coal from the belly of the Midwest to mix with the iron ore of the upper Great Lakes. The resulting steel would build for Cleveland the superstructure of a future far beyond that promised by the hulking loads of wheat and lumber floating down the canal at four miles per hour. It was too bad about the canal but, well, a city with the destiny of Cleveland couldn't be expected to subjugate its dreams to the rheumy reminiscences of old canallers. At Peninsula, the railroad demonstrated a similar impatience. Where the canal had gently lifted its body over the ninety-degree turn in the Cuyahoga River, the railroad simply slashed at the river itself, cutting a new channel across the neck of land that had given Peninsula the logic for its name. The railroad only understood the kind of logic dictating that a straight line was the shortest distance between two points, and that fast was better than slow.

Things have a way of choking on their own logic. Long before Jesus talked of reaping sown seed and perishing by the same swords that bring victory, God had let loose the principle that people must ultimately be measured by the values they impose on the world around them. So, too, with the railroads. The Valley Railroad could not withstand the irrepressible advances of the Baltimore and Ohio Railroad any more than the canal could stave off the Valley Railroad's iron thrusts. Despite Cleveland's resolution to keep the railroad out of the hands of the corporate giants, even if it meant narrowing the track gauge to a nonstandard width, the Valley Railroad became the property of the B & O ten short years after the first train whistled its way into Cleveland in 1880. The B & O bought more time—the better part of a century—but it, too, went the way of all who play "King of the Mountain" in business, leaving behind a series of sleepy structures lost in the weed-choked fields of Jaite where oil-blackened rails faded to rusted red. In downtown Cleveland, the jackknifed lift bridge that once threw the tracks over the Cuyahoga at the site of old lock #44 rusted to a stop in the up position, its arm pointed permanently toward the sky in futile supplication to corporate gods who never look down or back. Ironically, the old Valley Railroad has had a bit of the last laugh. The Cuyahoga Valley Scenic Railroad now operates passenger service for tourists traveling the heartland of the Cuyahoga Valley National Park. Occasionally, the cars are pulled by a visiting steam engine whistling a tune of a former glory.

That may seem a poor substitute for real railroading, but at least it involves trains and tracks. The Penn Central bed a mile to the east is now a much appreciated Metro Parks bike and hike trail. Persistent sprays of wild strawberries and an occasional apple tree (and, of course, the requisite patches of poison ivy) lightly laugh at the memory of the corporate monster that once lived here, perhaps recalling Penn's fiscal collapse that pushed the nation's understanding of capitalism to its limits when bankruptcy loomed in 1970. There was no romance in the high level discussions about bailing out the Penn Central. If it was to be done it would be done for the sake of the U.S. economy, and not to shore up teary-eyed memories of nineteenth century railroading. The knife that the railroads used to kill the canals had two edges.

Perhaps nowhere in the country is the rise and fall of the railroads more clearly demonstrated than in Atlanta. That city far from the Western Reserve came into the world with the unimaginative name of "Terminus" for no better reason than that it began serving such a function when a railroad surveyor drove a stake into open ground just west of Decatur (where the cautious citizens had spurned the proffered evils associated with becoming a railroad town). As anticipated, Atlanta exploded into a major American city during the short period separating its birth and the Civil War, riding the iron rails to fame where geography seemed otherwise disposed. The trains brought Atlanta many things, including the cold wrath of General Sherman who left his iron "neckties"—rails heated in the middle and twisted around trees—throughout Georgia and South Carolina. The general knew the importance of railroads. After the war, the yet-young industry had enough energy and power to rebuild Atlanta into the premier city of the South on the fresh ashes of Sherman's mixed blessings. But by the turn of the century, the city's people were growing impatient with their cumbersome iron benefactor. Atlanta's newspapers complained that it was noisy, dirty, and dangerous. Pedestrians unwilling to waste the five or ten minutes needed to wait out a passing downtown train would scurry across the tracks just ahead of the screaming engine. With increasing frequency, those dashes were taking people much farther than they wanted to go. The community answer was to begin constructing a series of viaducts over the main intersections, the end result of which was the eventual banishing of Atlanta's first story street level—and the railroads—to the city basement. People hearing about "Underground Atlanta" tend to imagine some Civil War era cubby-hole that had been rediscovered. In fact, it was originally an out-of-sight, out-of-mind cellar guestroom for an aging family member whom the youngsters had come to see as a nuisance and a bore.

It was the ending to a script the railroads themselves could have written.

~~~

It was the tracks, not the trains, that defined the parameters of our lives in Northfield. At Route 82, as the two lines squeezed toward each other to begin their final approach to Cleveland, they more or less formed the east-west borders for the Marshall Estate, that vast two-thousand-acre tract that gave us a wilderness playground as kids, and a good chunk of a national park later on. When we talked of "the railroad tracks" we always meant the New York (later Penn) Central line to the east. Miller's Hill, where we rode sleds and toboggans in the winter, was on the other side of the tracks. To get there, we had to either go up over the rail bed or walk hunched over through a storm sewer pipe that carried the spring-fed beginnings of Jensik's Creek under the track right-of-way. Northern Ohio's best blackberries were also geographically identified in terms of the railroad tracks, although by that designation we actually meant the huge open fields just west of the tracks.

Of the trains themselves, we saw little. No matter. They were just passing through, their sound giving us the only thing we ever wanted from them. The tracks stayed with us. While it was an article of faith that we waved to the caboose men in passing trains, there was little else friendly about the big diesel processions. At an early age, and with a good deal of reinforcement from Mom and Dad, we perceived the trains as dangerous. One smoggy Cleveland day, when I was no more than two or three, Mom took me to the bridge near our Beman Avenue home in the city's southeast side to look down at a train that had jumped the tracks. The engine, an old coal burner, big and black, was furiously spinning its wheels like a wild animal caught in a leg hold trap. The thing did not look friendly, and I had no desire to go closer. In the sooty railroad gully, which ran into the smelly, flame-shooting heart of Cleveland's steel-making district, the train seemed much at home. There were no pastoral greens or strawberries to round off the hard edges of this cold, mechanical thing.

Years later, as we were warming up for one of our Solon Foundry softball games at Cyril Bath field, I was reminded of that angry image from my childhood. Very deep in left field, far beyond anyone's batting range (except a heavy-set guy nicknamed "Poke" who, during Dad's fast-pitch playing days, wowed us with some stupendous blasts) Cyril ran a siding off of the main tracks to some of their warehouses and outlying buildings. Occasionally a single engine ferrying one or two boxcars would ease along the track and cross the road that ran by the ball field. On this evening either the engineer forgot to blow his horn or the approaching motorcyclist decided to beat the creeping train. At the very last moment he tried to turn away, but by then he was into the gravel by the tracks and spinning into a hopeless slide. His fall flopped him right into the middle of the tracks. As the train rolled over him I saw his arm and leg apparently sheared off by the screeching wheels as the engine came to a stop. Actually, he had managed to pull them in under the engine just in time, but the effect of that movement made it appear as if the limbs had been sliced off as easily as pruning shears take off branches. Mike Grubich, our manager, was under the train with the kid in what seemed like seconds. Knowing Mike, he probably had a good deal of experience with trains and their accidents. Not me. Like everyone else I had instinctively bolted toward the scene of the accident, but halfway there I wondered why. The hulking, hissing engine sat there menacingly, aware of its enormous strength and weight, defying any one of the agitated swarm of tiny humans to do anything about the wriggling creature it had trapped beneath its steel paw. Remarkably, the motorcyclist came out with just a few scratches. The number of his nightmares, I could not say.

 An old story from my childhood has it that Grandpa Knowles once told one of his grandchildren to crawl under a stationary train to get to some berries on the other side. I don't remember from whom I heard it, except that it was someone in the family. I suppose it could be apocryphal, but it has a ring of truth to it, especially the part about the berries. Grandpa went after berries with something akin to religious zeal, a trait that got passed on to Lloyd, Jack, and me. The railroad tracks bordered a field of blackberry patches that could yield us pickers a gallon an hour. Steamy August days would find us wading into shoulder-high weeds and razor-sharp thorns to mine the glistening black fruits that grew big as our thumbs. We filled the brightly colored metal cups that our Depression-minded parents thought suitable for children, then emptied them into a three-quart pot of Mom's. The only rule we abided was the "My bush!" claim for dibs. Each picker had his own style. Jack's was to eat a berry for every one he saved. Lloyd liked to follow along behind me and demonstrate that he could do just as well or better by getting the berries I left behind in my perpetual "grass is greener" eagerness to get on to the next bush. It was like Lloyd to make a science out of something I considered an art. To this day he teases me about eating corn-on-the-cob the same way, bouncing along to likely looking spots while he carefully excises each kernel. Sometimes I think he gets more enjoyment out of looking at the finished cob than from eating the corn.

More than the rest of us, Lloyd inherited Grandpa's sense for the significance of nature, in general, and berry-picking, in particular. Because black raspberries were in much more limited supply than blackberries, Lloyd occasionally determined that we should probably leave Jack out of the hunt so as to increase our yield. He would have us sneak out of the house and head off down Glencrest Drive in the opposite direction from the location of the "secret patch" which, magnifying our bad grace, was located on Uncle Frank's property about seventy-five feet from Jack's bedroom window. Nor was it beneath our dignity to pick berries from someone's property, and then try to sell them to the property owner. Once, however, we pushed our luck too far when we got into a lady's clearly cultivated and very unwild strawberries. When Mom made us take the obvious contraband back to the owner's door, the lady smiled tightly, took the berries, thanked us, and closed the door. It was then I began to suspect that not everyone in the 1950s was weaned on *Ozzie and Harriet* and *Lassie.*

But Grandpa would have approved the motive if not the method. Berries were important, and besides, his own sense of ownership wasn't always clearly defined, believing, as he devoutly and literally did, that God owns everything; we are just the stewards. And who is to say where stewardship began and ended? Our living room was, for years, adorned with an anonymous watercolor that he picked up from somebody's front yard long years ago, whether the product of a legitimate trash pile or someone's unfortunate timing on a home move I don't know. In that way, Grandpa was like the railroads with which I sometimes associate his memory, moving through life without focusing very well on anything except God's promised reward at the end of a single-gauged track. He never really fit into many corners of the society through which he passed. Keeping a job was difficult—even during the war—because fellow workers used profanity and smoked. Entertainment was almost always tainted with immorality, even playing with the "picture cards of the devil." Church was alright, but this grudging acceptance was rather severely limited to the Miles Avenue Church of Christ in Cleveland. He did allow himself the pleasure of the Cleveland Indians radio broadcasts, but after 1954 few people considered that much of a pleasure. Late in life he developed a fondness for—of all things—pool. He was living with Uncle Frank's family at the time

(Grandpa spent much of his adult life living with one or another of his seven children), and we were pleasantly shocked to see his agreeable interest in Uncle Frank's new purchase. He said he just liked to see the ways the balls sprayed out during a shot, and I guess that was true enough. There are few things as competitively aggravating to a seventeen- year-old kid as being beaten on three straight slop shots by his seventy-eight-year old grandfather who doesn't know a cue ball from a side pocket. But Grandpa was careful to tell everyone within earshot that he never played for money. No one watching him play ever doubted the soundness of that policy.

If Errett Knowles were my age today I suppose the world would be more sympathetic to him than was his world of 1887 to 1974. His wife died in 1929 after delivering their seventh child, on the heels of which came the Depression. Too, he may have been carrying around some sense of guilt or inadequacy for not having entered the ministry, as did his grandfather, father, his wife's step-father, two brothers, and a son. But there was no one to plead his case back then. Like everyone else, he took his wounds and went on. There wasn't much point in talking about them, and I don't recall that he and I ever did. To be sure, the wounds took their toll, the worst of which may have been on his sexual being. Grandpa never remarried, nor would his rigid religious discipline permit him any other outlets. The half century after his wife's death seemed to be filled with the alternating torments of desire and disgust. I don't believe he ever found an adequate storage closet for the cruel devil, but neither did he let it freely roam the house. As Mom once said simply, he should have remarried.

But even without these calamities, Grandpa was a bit of an ill-fit. Deep in my heart I suspect that most Knowleses are. The family tree is rife with branches wandering off in curious directions. Our people can be found in the heresy and witchcraft trials of seventeenth century New England (in the former instance as plaintiff, not defendant), among Richard's Crusaders in the Holy Land, and Edward's brutal warriors during the Hundred Years' War in France, tied by distant blood to Roger Williams and Benedict Arnold of Rhode Island, and stitched into the Bayeux Tapestry. We were Normans, debtors, heretics, slavers, and probably thieves, acquainted with kings and presidents and other lesser actors of the genealogical theater. Yet, through it all, there is the sense that we never quite put the other foot on the ground, that we never got fully in step with the human parade across the planet. Among the few leavings from Grandpa's life is a letter written to him by his wife shortly after Dad was born asking him to come home and name his child. I don't know where he was or why he had left—it certainly wasn't marital separation or desertion—he just seems to have drifted off for a life's moment or two.

This sense of drifting, of restlessly moving on down the tracks lest the train become derailed in some city of satisfaction or compromise, is of growing interest to me as I peek over the wall into the formerly forbidden land of old age. How much of the tendency is hereditary? How much of it is environmental? How much is simply and dutifully passed along by example from one generation of Knowleses to the next? And does all of this speak to a weakness in the world at large, a weakness in our bloodline, or is it an issue of weakness at all?

From what I can tell, Great Grandpa T.B. Knowles, Errett's father, also did little to support himself later in life. He seems to have spent the better part of his last few decades in the upstairs study of a paid-for house, supported by a wife who had raised nine sons, had several grandchildren living in her home at the time, and, according to family folklore, became

Cleveland's first female chiropodist (podiatrist). Meanwhile, T.B. was endlessly writing stuff destined to be forgotten. Nor was he the ever-present, kindly grandfather type, but rather seemed annoyed by the noisy brood that interrupted his scholarly deliberations. On one occasion, when Uncle Frank was about twelve and had dared to climb the forbidden tree in the yard, T.B. stormed outside and turned the hose on the would-be delinquent, the only likely result of which was a calamitous fall. Luckily, a gutsy neighbor girl, not too much older than Uncle Frank, intervened by disarming the old man on the spot. Apparently, T.B. seldom invested that kind of enthusiasm in a paying job.

Neither did his son, Grandpa Errett. I guess he did some landscaping in his time, and there was an unlikely misadventure as a front man for a con-man of sorts back in the early years of the century. (He related this fascinating story to me when we shared my dorm room the night before Lloyd's graduation from Milligan College.) Uncle Frank exited the formal world of work in his late forties, Dad in his mid-fifties. So, too, did I. We all had dutifully provided for our families, but just did not care much about careers at that time of life when other men often dive back into their work with a desperate vigor. Uncle Frank walked away from the vice presidency of a small but lucrative steel tubing business. I eagerly retired at fifty-six after thirty years in an interesting and well-paid research career.

It would be easy to conclude that the family genes contain a particularly resilient twist of DNA that drives the Knowles men out to pasture too soon. But that would not explain the genetic contributions of the Knowles wives who also went before us, most of whom were remarkably strong from start to finish. T.B.'s reminiscences, while scarcely mentioning his father, are lavish in praise of his mother, Lydia, whom he portrayed in terms of great strength and godly constancy. And if you asked any of the older generation of Knowleses, they would tell you that it was T.B.'s wife, Cordelia Baldwin Knowles, not her renowned patriarchal husband, who provided the family glue that held three generations together through disastrous death and the Depression. So, too, it seems, with Errett's wife, Florence, who despite her early demise managed to generate a respect that her children imbued with a sense of awe. Clearly her death was the one the family could least afford.

It is equally tempting to infer that the Knowles men simply succumb to the bitterness that in old age is sweetened by an ample supply of lifelong wounds and injustices. If the sinful world cannot be licked, neither should it be joined, leaving the third alternative of remaining aloof, admittedly isolated, but also unsullied. No doubt there was some of that in the lives of T.B. and Grandpa. I confess I can feel those attractive waters lapping at some of my own unguarded shores. Each year, I find myself less tolerant of what goes on (and what doesn't) in the Public Square and Marketplace. But these are, at least in part, inevitable by-products of advancing years. They take on different idiosyncrasies in different families. As kids, we winked knowingly at each other when Grandpa went into one of his long prayers, or when he made his annual "I probably won't be around next year at this time" speech at Christmas, or when he predicted that the end of the world was near. Eighty years ago, T.B. predicted that Scriptures and world events "seem to portend that the end of the present age will be about 2000 A.D." Now I am intrigued by the increasing length of the prayers of my father, my brother, and myself. The kids probably wink behind our backs, at least metaphorically. Children always wink behind their parents' backs, then head out into the world to produce a new generation of winkers.

Yet, our familial drifting is something more than the routine playing out of genetics or the funny habits of old age. There must be a better reason why the train slows down as it approaches its destination. For all of our old age grumbling about the deterioration of the world, we older Knowleses never quite manage any actions to back up our sour words. Indeed, we love babies, games, and music, all of which are children of hope, not despair. We might momentarily wash our hands of the whole human mess by proclaiming that we're glad we don't have to raise children in the kind of world we have today, but then turn around and invest that saved worry in our grandchildren and great-grandchildren. Grandpa Knowles insisted on passing on to us certain qualities—gifts, really—that he must have believed we would need for the fray, hardly the act of someone who was giving up on himself or those he loved. Among those gifts were a love of music and family pride. (When informed of our engagement, Grandpa told Lezlee that the best thing I would ever give her was the Knowles family name. During my poorer moments she hints that he may have been right.) Less planned but of greater value was his unyielding insistence on putting his faith at the top of his priorities, a point not, I think, entirely lost on his grandchildren. To be sure, Grandpa almost certainly did it the wrong way, replacing the joy God intended with knotty lumps of judgmentalism and, probably, guilt, but he nevertheless was pointing us in the right direction.

Perhaps we aging Knowles men entertain some late hesitation, a cautious look backward along the tracks to see if we might have missed something or someone along the way. Not long before Grandpa died, we took him a remarkably good picture of his bride that Mom and Dad had reproduced for him. There in the heavy late afternoon shadows of the rest home he looked carefully at the picture, then broke into tears. I thought I heard in his sobs not only sorrow, but also disappointment, perhaps regret. Maybe he thought of the time he had to be called home to name his son. Maybe he thought about how he had responded to her death by scattering some of her children—including my 4 year-old future father—among distant relatives far away from familiar siblings and Cleveland's comforting home sounds. Perhaps he had already entered that final darkness at the end of our conscious life that is the beginning of real accountability. Tolstoy's character, Ivan Illych, wandered through one of the world's greatest pieces of literature before inevitably reaching the same place, a harrowing blackness between life and death—far worse than either—when every human being must bring his pitifully shallow store of lifeworks under the scrutiny of the Eternal God. Ivan Illych did not break through the bottom of the blackness until he exposed the phony triumphs of his life: the vicious stair-stepping up the ladder of petty bureaucratic jobs, the vanity invested in a bigger house, the fathering of a family for which he never really served as father. Peace came for Ivan only when he at last opened his eyes and saw the thing he had religiously avoided all of his life: his Creator.

Grandpa may have felt some of that near the end. In the lingering twilight of life, truth may take on some terrifying dimensions. But I think that Grandpa, in his own odd way, had been searching for that moment of truth throughout his entire life. In fact, I suspect that he may have begun his search much earlier than many men do, perhaps when his wife died, perhaps when the then-unidentified phenomenon we call mid-life crisis began crowding out his remaining youth. And, in my more hopeful moments, I wonder if this searching is at the heart of the drifting pattern of Knowles men when they hit their forties and fifties. This is more than just idle speculation. After decades of reading and hearing the words of some of the greatest men and women the world has produced, I am increasingly convinced that all of life's meaningful activity is centered around people's search for God and, conversely, that all of life's folly is to be found along the paths leading away from him. Many people, perhaps most, readily follow those latter paths. Ivan Illych, seemingly well-liked and successful, was really nothing more than a vain, petty, self-pitying little man whose death meant a promotion possibility for his work friends and an end to a disagreeable presence for his daughter. (When, in the pain of cancer he became increasingly irritable, the girl sniffed to her mother that it wasn't *her* fault that her father was sick.) He had willingly deceived himself into believing that the bright trinkets of his life and his time in history made a difference, but at death's door he was forced to admit that his kind of life would have been equally wasted a thousand years before or a thousand years hence. It was just another slight variation on one of the oldest lies in the human story. What a great cost for such a cheap deception.

I suspect that this terrible but necessary realization is the visitor that disrupts the tranquility of some men in later years. It is not the kind of guest you would invite into your house, bringing with it, as it tends to do, a wardrobe of crisis, anxiety, and perhaps even severe depression. But pain, real pain, is a great inducement to getting us into the hunt for God. Pain also begets honesty. Amidst the carping of the over-fifty Knowleses about the deteriorating morality of the world and the virtues of the earlier days, I can hear the hound of heaven relentlessly asking, "Yes, but what about *your own* contributions to this failing world? And what are *you* contributing to these present, troubled times?" Many mid-life men run from such a voice. They seek the silliness of sports cars, young women, and fashions that advertise their foolishness to the world. Others become workaholics, or dive into completely new careers—anything to keep from having to listen to that voice calling them to accountability, the voice that skewered Ivan Illych to his death bed. The response of the Knowles men seems to be that of drifting, a general floating away from those things that once seemed so necessary, and toward some quieter corner—a berry patch by a railroad track, perhaps—where God can be approached on more agreeable terms. As I look at the world around me, the "drifting" seems not such a bad response after all.

I suppose I, too, am destined for an ill fit. The clothing already feels a bit uncomfortable.

∼∼∼

There were other trains and other tracks in earlier days. But these lacked the scope and grandeur of the big lines, instead satisfying themselves with smaller objectives and shorter distances. The Inter-Urban was a turn-of-the-century line between Cleveland and Akron that ran through Northfield down Route 8. At some point it was moved off of that main road through town a mile or so to the east, previewing the move that Route 8 would itself make to the same location when it bypassed the town in later years. The Inter-Urban was an important piece of transportation in its day, but of course the automobile changed all of that. Interestingly, however, the idea has come back around, largely because those same cars that liberated us are now choking our inadequate highways and airways in the urban areas. On the Eastern Seaboard the Baltimore-to-Washington train is as much a part of rush-hour as the jammed highways, while in Ohio, the legislature seems to be forever considering some kind of inter-urban rail system to link the state's closely clustered cities.

It is sometimes embarrassing to admit that I can remember the streetcars in Cleveland. Remembering streetcars is one of those era-identifying remembrances which, like remembering radio before television, or breadwinners walking to work, puts you way out of touch with a huge number of people walking around the planet these days. Nevertheless, I admit to remembering streetcars, even with the qualifier that it was the remembrance of a four-year-old boy. They were a fairly impressive sight to those young eyes, with electrical sparks popping from the end of the long rod that life-lined the car to the overhead wires. Long after the cars were pulled off the streets you could still find some of the tracks creasing the center of city streets. But no one would make bike trails there, for those tracks were servants, not masters, of the right-of-ways in Cleveland. Unlike the railroad tracks, those of the streetcars did not define the value of the land over which they passed. There were other important functions already going on, and the streetcars lasted only so long as they effectively served them.

There is no great body of folklore about streetcars. Songs about them are anecdotal and light, even funny, but seldom imbued with the sense of romantic glory once reserved for canals, railroads, and the other great, straight transportation ways of the nation. The railroads were their own excuse for existence, consuming the passions and fortunes of some of the country's greatest people. Not so the streetcar lines. While the lines did manage to get themselves tangled up in the robust, early twentieth century politics of urban corruption and reform they were never nation-makers. No one ever drove a golden spike into the final rail of a city traction line. Their job was to snake in and out of the city's corners in order to take people across street lines rather than state lines.

Now that both types of tracks have left their best days behind them, I wonder which ones were headed in the truer direction. We generally assume that bigger is better, so that part of the question has to be answered on the side of the iron horse. Beyond that the evidence gets kind of thin. A century ago, no one would have dared suggest that James Hill's railroad fortune was of no greater importance than the pittance brought home on a five-cent streetcar ride by a Chicago immigrant working in the Stockyards. Hill is now every bit as dead as that nameless immigrant, and we are beginning to suspect that the latter did more to build this country than the former. One man traveled the continent, the other barely traveled his neighborhood. One pursued wealth and power, the other pursued life.

I suppose our lives were meant to be trolleys rather than trains. We would like to believe that the tracks are long and straight, and that our destiny lies at the end of some singular path. But life is a series of starts and stops, sharp turns, dead ends, and crossing tracks. The people who do it best are those who quickly get started again. Maybe that is why our wives and mothers and daughters seem better at life than we are. They seem to know instinctively that God is not to be found at the end of transcontinental tracks, but rather along the short lines of the world.

Notes

1. Miller, ed., et al., *Independence*, 100.

2. Ellis, *The Cuyahoga*, 200.

3. Ellis, *The Cuyahoga*, 204.

4. T.B. Knowles, *Early History*, unpublished, 71.

5
Owners of the Clouds
The Indians

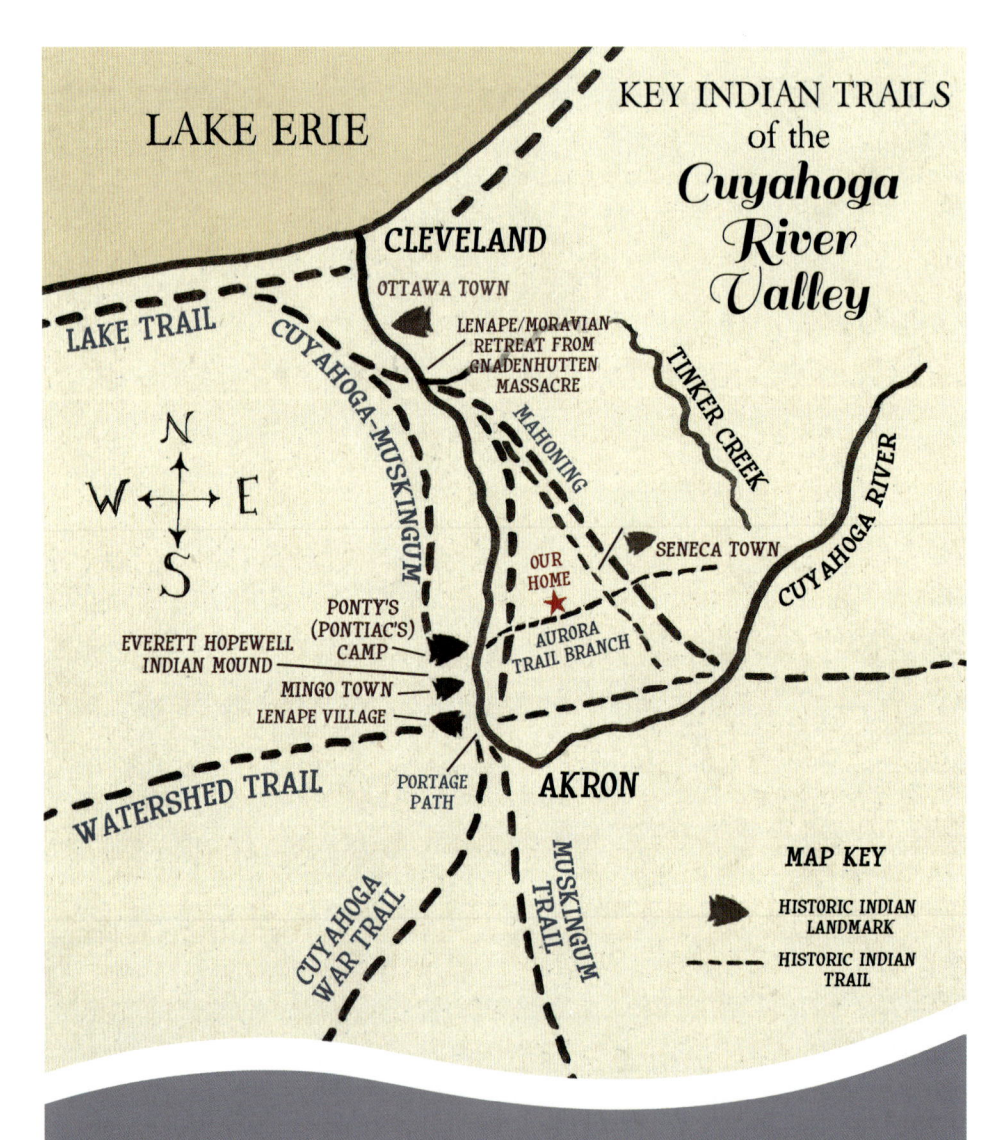

Indian trails map. Frank Wilcox's landmark map of the Ohio Indian trails, combined with the logic of geography, hint that a branch of a trail probably ran within a stone's throw of what would become our Northfield homestead.

Map image by Kristin Calhoun, based on Frank Wilcox's Ohio Indian Trails, Kent State University Press.

The real history of the "Western" Indians • The ones we almost knew •
The old trails of those who never owned the land, but were ever on it •
The spirit still occasionally finds a home

Centuries before America's founders cast covetous eyes on the continent's rich interior, the people indigenous to the land were regularly traveling the ancient routes criss-crossing the Ohio lands. The trails were hard won from forests so densely carpeted by the stifling hardwoods that even the Indians sometimes seemed intimidated by that jungle (to this day, Cleveland wears the moniker, "the forest city"). Not surprisingly, rivers were welcome interruptions in that forbidding, foreboding land, none more so than the Cuyahoga River. That river falls some four hundred feet between modern Akron and Cleveland, ensuring a good current for rapid transportation of the Indians' well managed canoes—and bringing untold numbers of these native travelers to within earshot of our future property in Northfield. Pontiac, Tecumseh, Logan, Little Turtle, they were no more likely to miss this crucial artery to Lake Erie and its shoreline trail than we would ignore an interstate route near to our homes.

The Indian nations that had stewarded, invaded, or simply roamed the Cuyahoga Valley for generations before the great white encroachment—Wyandots, Delawares (Lenape Nation), Ottawas, Mingos, Iroquois, a few wandering Shawnees, perhaps the Kickapoos, maybe the Eries—were far off my young path in the 1950s. As autonomous Indian nations, they had been largely disbursed (the Eries virtually extincted) since shortly after the War of 1812 when the Eastern Indian nations chose the wrong side for the last time. Tribal descendants were (and are) still there. Old Mrs. Hadlock across the street from us was full-blooded, or close to it. How I wish I had found out more about her! Virginia Chase Bloetscher states that some 5,000 Indians from ninety-five tribes were to be found in the Akron area in the mid 1980s[1]. But they are not the same presence that once saturated the valley. For the most part, I was wholly ignorant of the Native Americans nearby during my youth, except for the names that filtered down to us devoid of any cultural association: Chippewa Lake, Seneca Golf Course, Indian Creek, and—most meaningless of all—the Cleveland Indians. I had probably walked past and over scores of arrowheads lying openly in the shallow streams. But I wasn't alone. Ask most people in our neck of the woods, save the history purebreds, about Ottawas, Mingos, or Wyandot, and they might mumble something about the Boy Scouts. Mention the heyday of the American Indian and they will warm to the subject with what they know about the Little Big Horn, or the Apaches, or tumbleweeds and cacti, oblivious to the fact that the most important chapter in the white man's history of the Indians was written here, on the very ground under their feet.

But, as I said, I could only imagine the images of the indigenous residents in my mind. They did not leave enough of themselves for me to find. Unless you have the trained eye of an archeologist, like the renowned Charles Whittlesey of the nineteenth century, or are willing to look close enough at small rocks to see an occasional arrowhead, you will not see the original tenants at all. Considering that they had been padding on and paddling in the Cuyahoga Valley for a time period nearly fifty times the length of our brief stay here, their diminishment almost qualifies as a vanishing act. Sometimes, though, in the undulating ravines and thinning woods that form the eastern wall of our part of the Cuyahoga Valley, you half expect to see one. As a boy wandering those woods I occasionally would take a 360-degree scan in the hopes of just such a sight. At such moments the land seems to be remembering them. I would think to myself that some of them *ought* to be here.

Once, they were. It was a time when the great Indian nations still thrived, when their roles in the various wars with the white determined the fate of a continent. For a hundred years *after* the English began settling the New World—the length of time separating our present era from that of Woodrow Wilson—the region east of the Mississippi was dominated by the five tribes of the powerful Iroquois League. Over a century before Sitting Bull and Crazy Horse were merely prolonging the inevitable at the Little Big Horn, the great confederation of the Senecas, Cayugas, Onondagas, Oneidas, and Mohawks was helping to determine whether we would grow up speaking French or English. A few years later the great Ottawa chief, Pontiac, came close to determining the issue in favor of the French. Even as late as 1811 the visionary Shawnee warrior, Tecumseh, might have ended the life of the infant Republic's frontier had it not been for some extraordinarily bad luck. The point is that the Indians who regularly moved up and down the Cuyahoga Valley were of enormous importance in shaping the contour of our nation's history. The retrospective sense of tragic inevitability that we usually impose on our Indian history, the image of native peoples easily swept westward to the Pacific by unscrupulous and irresistible white forces, was little in evidence for most of the time white people were limited to this side of the continent; that is, roughly, 1500 to 1800. Things might have turned out quite differently with just the slightest editing of the script here or there.

The first of the great might-have-beens fell victim to the bad judgment of Samuel de Champlain. Early in the seventeenth century the French explorer decided to impress his Huron friends by teaching a lesson in manners to the warlike Iroquois to the south. Moving out from the recently established Quebec, Champlain's men had only to discharge a few volleys from their inaccurate but loud weapons to route the stunned enemy, leaving behind casualties that could be numbered on one hand.[2] But it was a Pyrrhic victory, costing the French an otherwise good chance for friendship with the one Indian confederation that could influence the continental fortunes of men, both Indian and white, once they accustomed themselves to the sounds of gunfire. Westward to the Mississippi and southward deep into Dixie, the Iroquois raiding parties were met with fear and tribute. From western New York they dominated the great events of America's interior, from the fur trade to the movements of tribes. The early misunderstanding with the French ruined a natural alliance with the Iroquois. Unlike the land-grabbing English colonists, the French trappers and traders were willing to live with the Indians on the latter's own terms with the land, taking what was necessary, giving what was needed, not caring that the man sleeping in that cave down by the river's edge was not a European. Nearly every tribe in the New World came to easy terms with the French, save for the humiliated Iroquois, and had the Five Nations (notably the Senecas) not thrown their weight behind the British and against the French at Ft. Niagara, or had they actually sided with their more natural allies, American history almost certainly would have been somewhat different. If Champlain had managed to stay neutral in the

Huron-Iroquois dispute, or had he chosen to see the Iroquois as trading partners rather than enemies, as was the wont of most of his countrymen quietly paddling the waters of North America, both the Iroquois and the French might have a more noticeable presence in this land. At the time, there was little about the destiny of North America that was inevitable. Who can say what might have become of the Indians had they had more time and space in which to develop a resistance to the incursions?

The will-o-the-wisp goes on. Had the French given Pontiac even a modicum of support for his fabulously orchestrated uprising, the British may have forever lost control of the interior, especially given the few years of American rule left to them at the time. And had Tecumseh not been burdened with an unstable brother, who was as maniacally destructive as the great Shawnee chief was gifted, the whole business might have come to a radically different ending when the Americans and British went to war just months later in 1812. But, of course, those things did not happen, and now the Indians are gone. But when they went they took with them the real Indian history of the great American "West."

~~~

The Iroquois certainly had a role to play in the Cuyahoga Valley. They were absentee landlords for a hundred years after the middle of the seventeenth century, and for half a century after that the few Senecas and Mingos who chose not to join the Five Nation Confederation dotted the landscape with their villages. A couple of miles to our east, Twinsburg was the site of a Seneca village, while Botzum, a few miles down river south of Boston, hosted the encampment of the famous Chief Logan, one of the few Indians whose words, delivered under the "Logan Elm," succeeded in finding a soft spot in the hearts of white men. But, on the whole, the great Five Nation Confederation left their mark on the valley by their intimidating influence more so than their actual presence. Indeed, the place seemed to have had a transitory effect on many of the Indians. Where a creek of any size joined the river, a village might sprout and flourish for a season or two, but would disappear just as quickly. Like the river itself, there was an undercurrent of residential restlessness in the valley.

A good deal of study has been devoted to the presence—or absence—of the Erie Nation in the valley. The Eries were an intoxicatingly mysterious, prehistoric tribe that left its name everywhere and its provable physical presence almost nowhere. Like the Ten Lost Tribes of Israel, the Eries seem to have vanished after their brutal defeat at the hands of the Iroquois in the 1650s. Some early scholars suggested that the Eries had peopled the heavy woodlands of northern Ohio, western Pennsylvania, and perhaps even western New York, for over five hundred years. During their long and agonizing defeat, many of them were, like the old Israelites, carried back to the Iroquois homes by those latter-day Assyrians, there to be tortured or incorporated into the families that had lost husbands, brothers or fathers in the endless Iroquois wars. The latter was a tradition among many of the Indian tribes, and they frequently made use of it when they took white prisoners, especially children. Eastern Indian history is filled with stories of white children raised by Indians who were none too happy about being returned to their blood families many years later. Unfortunately, torture was also a well established tradition, especially among the Iroquois. Years after the decisive Erie defeat—probably at their stronghold at Rique, which is variously identified as Conneaut, Ohio, or Erie, Pennsylvania—the Iroquois were still enjoying their burning of Erie prisoners. Like a cat toying with a dazed mouse long after the creature loses its capacity to provide hunting sport, the Iroquois would allow Erie boys to grow to manhood so that their fires might burn under them for a day, or, with luck, two before death mercifully interceded. Our contemporary talk of "roasts" and "slow burns" convey none of the anguish literally visited upon the Eries.

But assimilation and torture cannot explain the disappearance of 15,000 Eries. Neither is it likely that they simply moved to another part of the country and assumed or were given a new name, as were the Wyandots (Canadian Hurons) and Mingos. The nineteenth century Indian authority, Henry Schoolcraft, whose "authority" has not always been recognized by later historians, suggests that some of the Eries fled south to become the Catawbas, while others have hinted that the remnants migrated westward with some of their former Iroquois enemies to reservations in Oklahoma.[3] It is likely, however, that such mass movements would have left better tracks, even in the land the Europeans called the "black forest" three and a half centuries ago. The more probable explanation is that the Eries were simply extinguished by the Iroquois. The extremely proud and warlike people continued to fight their Iroquois conquerors for many years after the final outcome had been decided, falling back from one fortified encampment to another as they retreated stubbornly up the river valleys reaching down to Lake Erie. Lupold says only 600 ever surrendered. The descriptions of slaughter on the heels of Erie defeats, such as that at Rique, imply that at such moments the Iroquois were no more of a mind to take prisoners than were the Eries to voluntarily provide them. Women and children apparently shared the same fates as their husbands, brothers, and fathers.

The mystery of the Eries continues today, but in a way that hits closer to home. In recent decades, some historians have argued that the Erie empire has been overestimated, and that their western reaches never broached northeastern Ohio, much less the Cuyahoga Valley. Basing their case on archaeological and cartographical evidence (the "battle of the maps"), they claim that some other tribe, of Algonquian origin, perhaps the Kickapoo, was responsible for the scant Indian leavings in the valley. (see Phillip R. Shriver, "The Beaver Wars," and Jarer and Eric Cardinal, "Archaeology and History," in *Ohio's Western Reserve*) Lupold even goes so far as to directly reverse himself concerning his earlier assumption of Erie residence contained in his classic work on that tribe, *The Forgotten People*. Apparently the issue is to be relegated to the land of academic question marks, awash in a murky sea muddied by the work of bulldozers and plows, and pursued by a few scholars who are forever racing against time and new shopping malls.

There is something within me, stronger than my training as a researcher and historian that resists the notion that the Eries were never in the Cuyahoga Valley. To be sure, part of that something is emotional and not very easily defended. That is the part that says that the great tribe just *had* to be here. They are the ones I could feel in our woods, the real land owners who had been there before Columbus or Leif Ericson or whoever it was that first wandered ashore in America. Four times I have treated myself to William Donahue Ellis's *The Cuyahoga*, and four times I have been mesmerized by his tragically haunting description of the Eries' final retreat up the Cuyahoga Valley to Copley swamp and the battle that sent them to final oblivion. From the vantage point where my heart has placed me, high atop the old Indian lookout point across the valley floor from Red

Lock Hill, I can see the fiery panorama unfolding before me. It is night. The famous Erie signal fires are burning brightly, carrying the alarming message the length of the valley (in itself, not so fanciful, actually; Bloetscher states that Cleveland's Terminal Tower is visible from such a lookout point near Peninsula, over twenty miles south, and Ellis confirms the possibility given the reflection a large fire would make). Each of the walled forts sitting on top of the successive ravines, including the two Whittlesey excavated almost in our backyard between Red Lock Hill and Rt. 82, become scenes for savage fights as the relentless Iroquois storm the walls under the protection of their overturned canoes. In each place, the conflict burns brightly for a historical moment—perhaps a day, or a month—like a lighted sparkler, then passes on to the next fort where more of the Eries have determined to stay either on the land or in it. The final siege at Copley swamp takes weeks, perhaps months, before the last Erie signal fire is forever extinguished. The Iroquois, utterly spent by the stupendous Erie resistance over the years, cannot even summon the energy to return to their homelands in New York, forcing them to winter here, licking their wounds and burying their numerous dead. When they finally exit in the spring, they leave behind them an Indian ghost town where the Iroquois don't care to live, and no other tribe dares to.

The evidence for an Erie presence in the Cuyahoga Valley is not all emotional. The traditions for that presence are long and not necessarily apocryphal. A few of the first reliable white observers in the American interior, the resolute and fearless Jesuits, placed the Eries here, as have some historians since.[4] The Jesuit priests among the Hurons and Iroquois wrote of the Eries on the eastern and southern side of the lake that would later bear their name. But the Jesuit fathers apparently seldom traveled to the Erie country during the tribe's existence there, and it is doubtful that white eyes ever saw the Eries in their strongholds. Even when a Jesuit or two broached the Erie territory, their message was not well received.[5] That is probably just as well. The Eries were primitive and warlike even by the standards of the Iroquois, whose blood they distantly shared. As the Erie warriors at Rique are reputed to have said, in response to the Iroquois taunt that the Jesuit Master of Life was fighting on their side, the Eries honored no gods except their bows, arrows, and hatchets.

The Iroquois may have won the final battle, but the evidence indicates that many, if not most, of the earlier ones were won by the Eries, especially those fought before the five New York tribes united into the Iroquois Confederacy in 1575. The western-most Senecas bore the brunt of the Erie enmity. Bloetscher relates the story told by Black Snake, an Indian chief believed to be more than 100 years old, who, in the wake of Perry's naval victory on Lake Erie during the War of 1812, told of a much earlier water engagement in which the heavier canoes of the invading Eries proved decisive in a crushing defeat inflicted on the Senecas in their smaller craft. The old chief concluded that the resulting slaughter in that battle taught the five tribes that they needed to unite with each other if they hoped to survive their powerful western neighbor.

The Erie prowess, combined with their tribal longevity in the southeastern Lake Erie region, make it unlikely that they would have ignored the strategic Cuyahoga corridor in their rear. The Cuyahoga-Tuscarawas-Muskingum river network that led to the Ohio River, with its eight-mile portage at Akron, was well known to the Indians in this part of the continent many centuries before Columbus landed. Furthermore, as already noted, the abundance of high ground along the often narrow Cuyahoga

Valley would have been a natural attraction to the fortress-minded Eries. It would have taken a very compelling reason to keep the Eries out of the Cuyahoga Valley, one much stronger than the Kickapoos of the milder Algonquian extraction. The history of the Eastern Indians in the United States is a huge jumble of movements, wars, and alliances in which the numerically superior Algonquian tribes struggled to avoid the militarily superior tribes of Iroquois blood. Frequently, the latter aided the former by warring among themselves, as in the Five Nation's wars against the Hurons and Eries, but the more passive Algonquians were seldom a match for the Iroquois tribes before 1700. It does not make sense that the Eries would have allowed a weak, Algonquian tribe to hold on to such a treasure as the Cuyahoga Valley in their own backyard. This is not to say that the entire Erie tribe ever lived there—far better historians than I have concluded otherwise—only to say that they must have been familiar with the place; perhaps familiar enough to fortify against western incursions.

The causes of the final Erie-Iroquois war are partially obscured by stories handed down via nineteenth century historians that are sometimes difficult to distinguish from legend.[6] Because the Eries were Ohio's last undocumented people, and later observers had good reason to favor the Iroquois case, many of these accounts seem slanted against the Eries. Most notorious is the story of an Erie challenge to the Senecas for a ballgame to prove tribal superiority. The account would have us believe that the Senecas were properly reluctant about responding to such an obnoxious stunt, but not at all reluctant about winning once they did. Thereafter, the infuriated Eries proceeded to further embarrass themselves by losing a foot race and wrestling match, whereupon they decided that valor was the better part of discretion and so planned a sneak attack on a Seneca village.

Better substantiated are accounts of a peace conference, perhaps one in which the Eries were offering some kind of tribute to their eastern neighbors after three quarters of a century of wars with Iroquois confederacy, during which a Seneca was killed by an Erie, probably accidentally. The Iroquois proceeded to slaughter all of the remaining members of the Erie peace party, save a couple whom they allowed to return home with the story of the outrage. By this time, the solidly united Iroquois tribes were confident of their ability to finish off the ancient, lingering hostilities between the two nations. Another perspective, also apparently grounded in at least some truth, depicts the Erie capture and torture of an important chief of the Onondagas, the central tribe of the Five Nations in whose longhouses the sacred council fire was kept. The Eries had followed the tradition of offering the chief to the sister of a slain Erie warrior in the confident expectation that she would accept him into her family as a replacement. Instead, to the consternation of the Erie tribesmen, she demanded her alternate right to see him tortured, adamantly refusing to listen to pleas to the contrary. As the fire was lighted under the chief, he is reported to have said that the Eries were lighting the fire that would consume their nation, a fairly accurate prediction, though one he never saw fulfilled.

The intertribal frictions of two proud and warlike peoples may have provided the sparks for the Erie-Iroquois war, but it was other larger issues that made such a conflict inevitable. The British, whose main Indian trump card in the New World was the Iroquois, were more than happy to see them at odds with the other Indians of the interior. Such agitation not only destabilized powerful and potentially obstructionist tribes like the Eries, but also hampered the otherwise unrestricted access that the French traders seemed to enjoy to the continent's vast interior. Even without English encouragement, the Iroquois were motivated by an increasingly desperate economic factor, that being their need for furs. Because their own hunting grounds had been pretty well played out in the feverish rush to supply the voracious white appetite for furs—a one-sided game in which the Indians would sometimes trade a bundle of furs stacked to the height of a rifle in exchange for the gun—the Iroquois coveted the richly pelted lands to their northwest and southwest, those being the lands of the Hurons and Neutrals in Canada, and the Eries in Pennsylvania and, possibly, Ohio. When the Eries became the last of the three to succumb in 1657, the Five Nations had virtually unrestricted access to all of the hunting grounds east of the Mississippi, except for the sliver of land being held by the stubborn but inconsequential English colonists along the Atlantic shores.

That is when the Cuyahoga Valley became an Indian ghost town, at least, for a century or so. For 150 years thereafter, the valley would not be important because of who lived there. But it would be crucial in terms of who passed through it. Most prominent among these was Pontiac, war chief of the Ottawas and architect of the greatest Indian alliance in history. Ellis states that, as a boy, Pontiac spent much time in the Ottawa camp that subsequently bore his name, ("Ponty's Camp") and later visited the place on trips up or down the valley.[7] The location of the camp was only a short distance from the foot of Red Lock Hill, on the west side of the river near Columbia Road. I am hopelessly fascinated by the almost certain prospect that this historical giant, whose famous "conspiracy" humiliated the British military with the quick capture of numerous forts that nearly cost the Crown its huge trans-Allegheny possessions, made the same boyhood romps up the ravine to the falls and Jensik's Creek as I did. He may have hid under the same large outcroppings, wondered at the same half-buried boulder in Uncle Frank's backyard, and chased the ancestors of the salamanders that so intrigued me as a boy.

In my more fanciful moments, I even wonder if he played ball on the ground that became our ball field two centuries later. The prospect is not altogether silly. Even in the heavily wooded Black Forest there had to be open fields somewhere. And there is no doubt that the Indians loved to play ball. They called the game baggataway, a violent forerunner to modern lacrosse. During my year as an Explorer Executive with the Boy Scouts in Georgia, my territory included a county north of Atlanta that was home to the small town of Ballground, Georgia. In trying to familiarize me with my new district one of the older executives told me that the Cherokees used to convene there for epic versions of the game, sometimes lasting several days, often involving several deaths and always many serious injuries. The problem was that the Indians were as likely to use the sticks to bludgeon each other as to advance the ball toward the opponent's goal. Such advancement could be accomplished by the sticks or virtually any part of the body other than the hands. Allan Eckert's matter-of-fact description of the game's objective sounds like Candyland and War on the same rule cards.

> The opponent does everything in his power to prevent his post from being touched by the ball. Tripping, hitting, clubbing, kicking, and other assorted forms of mayhem are permitted. Extreme roughness is standard procedure. If one of the Indians on the opposing team has the ball between his feet and is hopping with it, a defender has every right and

obligation to rush up and, with all his strength, strike the man with his racket on the feet or legs, sometimes on the body or head.[8]

Pontiac knew the game well. During his attempt to lull the British commander at Detroit into believing that his intentions were peaceful, he had his warriors engage in a game within view of the fort's walls. The British commander, forewarned, did not take the bait (at another fort, the trick worked to perfection when the soldiers eagerly streamed out to watch and wager the well-loved spectacle; after several hours of bloody sport, the Indians suddenly turned on their spectators and took the fort without difficulty). Later, when Pontiac rightly suspected that an old squaw converted by the Jesuits had tried to betray him, he grabbed a baggataway stick to administer the vicious beating.[9] It is possible that Pontiac's boyhood victory screams echoed across our ball field long before our more subdued shouts.

Somewhere shortly thereafter, the similarities between us end. As an adult warrior in his own village, he frequently went naked, so long as the weather obliged, exposing an array of strange tattoos. A white bone curled through his nose. During great war councils, he would frequently cover his entire body with war paints, usually outdoing all of the other warriors. Even more unusual was Pontiac's hair style, but he had a very good reason for it.

> His hair was trimmed in traditional Ottawa fashion—shaved almost to the skin at nape and crown, and then gradually tapering to a brush-like length of an inch or so high above his

forehead. Long ago he had learned from his father to cut his hair in this manner, so that an enemy could not grasp him by the hair with one hand and plunge a knife into him with the other. And more than once his own hand had clasped the hair of a Cherokee in just such a manner and jerked the enemy off balance while his other hand directed a knife or tomahawk to the Cherokee's breast or head.[10]

Pontiac would die with a hatchet buried in his own head, delivered by a fellow tribesman in the complicated chemistry following the chief's great conspiracy. His vision of driving the English from lands that were never the possession of the French to give away died precariously short of its fulfillment. But the idea of Indian unity in the face of white encroachments did not die.

A half-century after Pontiac came Tecumseh, the extraordinarily gifted Shawnee warrior who aimed to draw together most of the tribes east of the great river into a gigantic force that would hurl the whites back beyond the Appalachians once and for all. Eckert sees Tecumseh as a pragmatist who knew that the sign he mystically predicted to signal the uprising would sooner or later be provided by the law of averages (it came in the form of a black rain, an unusual but not unheard of phenomenon in these parts). But Eckert also seems persuaded by the strange visionary powers of Tecumseh. The great Shawnee ranged all over the Eastern United States, awing distant tribes with his spellbinding message of things to come. Unlike Pontiac, Tecumseh added very precise details to his prediction concerning the signal that would call the Indians together, much of which seemed physically impossible.

In the midst of the night the earth beneath would tremble and roar for a long period. Jugs would break, though there be no one near to touch them. Great trees would fall, though the air be windless. Streams would change their courses to run backwards, and lakes would be swallowed up into the earth and other lakes suddenly appear.[11]

But, in fact, the event he called for *did* happen. The greatest earthquake in the history of the Midwest, which tore the earth along the still ominous New Madrid fault line on December 16, 1811, sent the waters of the Mississippi River backwards for a short time. Lakes and streams appeared instantly on land that had been dry since the last glacier, while other waters were sucked into the earth's cracks leaving a confused chorus of flopping fish and squawking mallards. Forests fell, livestock were pitched drunkenly to the swelling ground, and cabins were turned to matchwood. Eckert appears convinced that Tecumseh's powers to foresee such an event qualify him as one of the great prophets in the nation's history. "This was the earthquake that occurred where no tremor had ever been recorded before; where there was no scientific explanation for such a thing happening; where no one could possibly have anticipated or predicted that an earthquake would happen. No one except Tecumseh."[12]

Unfortunately for Tecumseh his great strength—his capacity to inspire disparate Indians who were separated by old enmities and hundreds of miles—proved to be his undoing. While thus evangelizing far away from his great encampment of Indians at Tippecanoe, Tecumseh's unstable brother, The Prophet, who had only enough of his brother's gifts to be dangerous, led a rash attack that wasted the confederation's power and sent William Henry Harrison on his road to the White House. Tecumseh would die far from his native Ohio as a British confederate—some say as a brigadier general—at the Battle of Thames during the War of 1812.

Indians were not in the habit of keeping diaries or drawing maps, so we know little about the details of Tecumseh's extraordinary travels. However, it is a fair bet that he was no stranger to the Cuyahoga Valley. Not only was the valley a main Indian travel route to the northeast, it was a melting pot of Indian bands and villages that this war missionary was not likely to have overlooked. In the waning twilight of Iroquois power, groups of Ottawas, Mingos, Senecas, Delawares, and others had taken up uneasy residence in the coveted land. The chains and markers of Moses Cleaveland's 1796 surveying party alerted those Indians to the fact that the English-speaking people obviously intended something more than hunting and trapping expeditions. That they were ready listeners for Tecumseh's appeal is evidenced by the tremendous sense of expectancy that gripped the tribes on the eve of the War of 1812. His canoe, like Pontiac's before him and the canal boats of a later day, graced the waters of the valley. It is a fair bet that his feet knew the ground that was Ponty's Camp and became the town of Jaite at the foot of Redlock Hill.

~~~

Pontiac, Logan, Tecumseh—they were all here in one way or another, although in typical Indian fashion they left little tangible to show for it. I think the Eries were here, too, though perhaps only as visitors, but they left even less of themselves. Not even the oral traditions or the written records of white men can completely verify their residence. There is, however, one exception to the otherwise dearth of Indian leavings: arrowheads.

Concerning Fort Island west of Akron, in the midst of Copley swamp near to where William Ellis believes the Eries made their last stand after retreating stubbornly up the valley, Virginia Chase Bloetscher offers this tantalizing footnote: "A long-time Akron resident recently recalled that in his boyhood days, he and his friend gathered up bushel basketfuls of arrowheads, using rakes, from the surface of Ft. Island. Thus, it would seem that a great battle had been fought there at least sometime in history."[13]

Funny. For decades my only association with the town of Copley was my memory of a high school football rival. They grew huge farm boys down there who made intimidating football players on teams that usually fared better than our Nordonia teams in the race for Suburban League championships. Had the Eries been a people of recorded history, or had they survived as a tribe, Copley might now be a shrine of sorts for them. If, in fact, the arrowheads are of Erie origin, they serve as an entirely appropriate testimony to the great warrior tribe. The Eries were famous and feared for their use of poison arrows. After dipping the arrowheads into the poisonous innards of certain animals, the Erie marksmen could dispatch eight to ten arrows in the time it took the Iroquois to discharge and reload one shot from their guns. Ultimately, however, those guns proved decisive in the hands of the Iroquois, just as they had many years before in the hands of Champlain's men when used against the same Iroquois. But those bushels of arrowheads at Ft. Island tell us that such battles were not decided for want of courage or effort. If they were there, it was a fitting way for the Eries to end their centuries-long rule of the eastern end of the lake that would bear their name.

Lloyd used to have an arrowhead collection. He may, in fact, still have it. Lloyd always was a collector, of butterflies or shells or whatever else he was appreciating in nature at the moment. He was always much more patient than me in looking for such things. I could only see an infinite number of rocks, but Lloyd saw quartz, micas, and arrowheads. Once, in our forties, when I was relaying some bit of trivia to him about the Erie use of arrowheads made poisonous by dipping them in animal livers, his eyes brightened. He proceeded to explain that one of the arrowheads in his collection had a reddish hue to it, and that he always suspected that the stain was poisonous.

If anyone has a right to a piece of the illusive Eries it is Lloyd. More than any of the rest of us he inherited what little remained of the Indian spirit that drifted in thin wisps across our land in the 1950s. He didn't have to invent his Indian fantasies, but seemed a kindred spirit from the moment he set foot in Sagamore Hills. That is not to say that Lloyd knew much about them. He would study a great deal of history on his way to two masters degrees and a Ph.D., but most of it was concentrated elsewhere. He probably knows considerably less about our native tribes than I do, and like most of Northfield's residents, probably has little idea of the enormous importance of this part of the country relative to Indian history. But Lloyd didn't have to know about the Indians who walked his land, for I am convinced he could feel them.

That feeling began with a sense of sacredness about the land and the creatures that walked, crawled, and swam upon it. It is said that for many generations the Indians would not take up permanent residence in what is now Kentucky because they considered it a sacred hunting ground. Its beauty and abundance of game, rather than fostering greed and the covetousness that begins and ends in the sanctuary of ownership, apparently convinced the Indians that here was land that was too precious to become the property of any person or tribe. They traveled to it, took their kills, then retreated so as not to despoil the holy ground with their moccasined feet. I think that Lloyd always had that same sense of other-worldly esteem for the vast wooded area that surrounded our home. Whenever anyone special came momentarily into his life—a school friend in his boyhood days, or later, a girlfriend—he would always take them back to the ball field to show them the woods and falls. I am not sure what he said at such times; it probably seemed awkward and unusual, but I think he was trying to say that, "This is a big part of me. You can't understand me unless you first understand this." Not everyone did. Once he brought a girl home whom he seemed to like very much. He thought she looked like Karen Carpenter, and appeared to possess only one fault; she smoked. One summer evening they were having some kind of disagreement outside (I think they were beginning to realize that this one was not going to work out) when she angrily tossed her cigarette into the grass. Lloyd probably scotched any chance of redeeming the relationship when he proceeded to go over and pick it up. She could not be blamed for concluding that he cared more for nature's feelings than her own. He did.

That was vintage Lloyd. The sight of an old tire in Jensik's Creek or a freshly cut dirt road into a new housing development could bring his blood to a boil, but the routine stash of human debris cluttering our property—clothes poles, ball gloves, and blackened pieces of half burned newspaper—seemed not to bother him at all. It was as if he were honoring some ancient agreement that said that the land does not begrudge you a corner of space if you are willing to otherwise live in harmony with it.

The Indians practiced that same harmony. When they hunted, they killed only what they needed, used virtually every part of the kill to fill some need in their lives, then they took a moment over the dead animal to apologize for having to do such a thing and to thank it for its sacrifice. The idea of killing animals for sport, or simply for a tail or horn, was utterly foreign to them. Lloyd felt that sense of alienation one summer day long ago. It was one of those days when there is nothing to do, when kicking a stone down the road or watching an ant drag a dead wasp through a field can cost a kid two hours. Lloyd had made a bow of sorts, maybe one from a kit, but perhaps just a handy stick that he knife-notched at both ends to accommodate an old piece of string. Whatever the quality of the bow, it was almost certainly better than the arrows. These Lloyd had made by skinning the thin shafts of dried goldenrod stalks. He knew that they were too light and too crooked to ever hit anything, or do any damage if they did, but on a slow summer day anything will do. Absently he aimed the unlikely weapon at a frog sunning itself by the large, water-filled hole that had been left when Dave gave up on his fort. The arrow zipped through the steamy air and skewered the frog to the wall of the hole. I can almost hear Lloyd tearfully apologizing to that frog. I have a suspicion that he feels a bit badly about it to this day. For any other kid it would have been a normal exercise in curiosity; for Lloyd it was a violation of a sacred trust.

There were times when Lloyd's love of nature got him into trouble. Our Aunt Doris was one of the most loving creatures God ever created, and always a great friend to kids. But she got pushed past even her limits when Lloyd assumed she would share his enthusiasm for the snake that he had captured and brought into her house. Lloyd's first problem was that his idea of "capture," based as it was on the principle that you not only never hurt a living creature but shouldn't inconvenience it much either, was to allow the snake to drape itself over a stick. His second problem, and the beginning of a very big one for Aunt Doris, was that the snake apparently got bored in that position. Off he went in the middle of Aunt Doris's living room, never to be heard from again. I don't think Aunt Doris stopped thinking about that snake until the house burned down many years later, long after they had sold it.

Nature is still one of Lloyd's causes, and still one that could get him into trouble. Some years ago, on the road that runs past his home in Lansing, Michigan, he saw a turtle making its slow way across the dangerous street. He started out after it, but a car was coming. As the turtle and car were on opposite sides of the road, Lloyd assumed he would be able to effect his rescue after the car passed. But a few feet from the creature, the car deliberately swerved across the road and crushed the turtle under its speeding tires, providing a momentary laugh for a car full of teenage kids. Lloyd raced out into the road behind them, yelling and shaking his fist, sprinting a few yards in their direction in the hopes that they would take the bait and come after this more interesting game. They did not, but I wonder at such a scene: an ordained minister and bible college professor in his forties, who to my knowledge has never been in an adult fight in his life, piling into a swarm of teenage guys, who are probably half drunk, to avenge the death of a turtle. Most people wouldn't understand it. An Erie warrior would have.

I think it was the Indian shadow over Lloyd that made him something of a collector, observer, and chronicler. His collections were his ultimate compliment to the world around him and as close as he would come to attempting to own a piece of it. They were microcosms, miniature re-creations of that larger world, and testimonies to the way he wanted it to stay. He became a chronicler of that world long before he became a historian, concerned with the whats before he pursued the whys. The habits spilled over into other parts of his life. Every night until long after he was grown, Lloyd would record his bedtime (I think he still does). Did he ever look for patterns, or correlate his sleeping practices with the events and feelings of his life? I never asked him; probably not. And even if he had, those exercises would not have been his main reason for keeping the bedtime log. The information was important to him not because of what it did, but because of what it was: truth. Pure, simple, natural, factual truth.

The differences between us in this regard were readily apparent in the ways we played as boys. Our lives were filled with the usual amounts of marbles, baseball cards, and toy soldiers needed to define a boy of the '50s. Lloyd's idea for such games was to recreate, as near as humanly possible, the exact circumstances of real life. It would take him hours to play his "marble baseball" game because he had to set up each baseball card in its right position on the field—heaven forbid that first-baseman Tito Francona should have to be put in right field for want of a Rocky Colavito card—an agonizing exercise he had to go through each half inning. Then, of course, the at-bat team had to be put in its proper lineup. A marble was used for the ball, a popsicle stick the bat, and some provision had to be made for the crowd and announcers. I recall that he even kept a scorebook and some individual statistics. Lloyd used a whole room for this extravaganza, and at any time could tell you which team was in fourth place and whether Gary Bell was likely to be starting or coming in out of the bullpen. My baseball games, like all of my imaginary games of play, consisted of a single toy, one "little man," vigorously manipulated three inches in front of my nose by agitated fingers that made the little soldier or Indian or cowboy dance around in a series of arbitrary jerks and twists. That was it. Everything else played out on the stage of my imagination. I only needed the space taken by the chair I was in.

Lloyd's passion for naturalness, for seeing the world without the adulterating touch of man's hand, led him to a moral rigidity that would have been stifling in anyone else. Lies, for example, were wrong—not sometimes, usually, or virtually always, but *always*. Truth is natural and simple (not easy, but simple), and therefore right. This conclusion came not from the insecurity born of the fear of other options, not from blind adherence to some verse of scripture, not from any deontological imperative to do good, but simply from Lloyd's observation of moral truth in the created world around him. It is a view unaccompanied by rancor and rage. It is a garment he wears with grace, dignity, and humor. I am now much closer to that view, myself, than that which I embraced in younger years when I was sure that something as important as morality had to be more complicated than that.

We argued about it, once, loud and long. It was during one of our Christmas vacations while back home from Milligan College. The local school superintendent (Bill Boliantz father of my classmate, Patti) had called and asked if we wanted to make a little money by dismantling the bleachers at the former high school football field. In the midst of piles of steel tubing and long two-by-eights painted green, we hotly debated whether it would have been right for a Christian to lie to the Nazis to protect the life of a Jew he was hiding. The argument did not abate as we broke for lunch and headed for the Dairy Isle across the field. Fresh from Fletcher's *Situation Ethics*, I was sure that this extreme example fell within the bounds of acceptable—even required—moral behavior. Lloyd did not think that a principle could be held hostage to every situation that happened along. The argument was not solved at the time, a fate common to Knowles arguments, but later reflection has convinced me that I may have won the battle but lost the war.

Despite the particular positions we might occupy during an argument, we both tacitly understood who held the moral high ground when it came to important, practical matters. The point was demonstrated very clearly one Easter morning. We had both awakened with the sense of excitement that comes with a holiday and a competition, the latter in the form of our annual Easter egg hunt that took place first thing after we got up. Not unreasonably, I said that I had to go to the bathroom, a point that Lloyd had to concede, but one that made him nervous since the trip might afford me the opportunity to scout out a few of the less well concealed eggs and so better plan my search. The dilemma was solved a moment later when Mom and Dad saw Lloyd leading me to the bathroom, one hand on my arm, the other firmly over my eyes. The great testimony in that little episode was not the fact that Lloyd had seen fit to blind me, but rather that I never for a moment considered the need to make the same requirement of him. He was Lloyd. The integrity of the hunt was secure.

I suppose there is something universal about big brothers, but different age spreads alter that something. Lloyd and I are separated by only twenty months, and except for a year or so around the time we were fourteen and twelve, we have been about the same size all of our lives. This meant that for most of our young lives we competed on fairly even terms. He did not have to figure out ways to patronize my efforts or to get rid of me in order to find someone more worthy of his skills. My memories of him do not include a time when he beat up an older boy who was bullying me, or a day when he gave me his old three-speed bike so that he could graduate to a

with brother Jeff   4/10/49

Lloyd (2) and me (11 months). A classic image of my older brother's role in my life, with an eye looking ahead and a protective hand on my chest. *Photo from author's collection.*

car. At such times, he would have been equally at risk facing the bully, and equally interested in riding the three-speed. But such things are, I think, greatly overrated, especially in the entertainment industry. Brothers who are far enough separated chronologically to have had such experiences are also far enough separated to have missed out on hundreds of others. Given the choice between a sometime champion and a fulltime companion, I am glad I got the latter.

No doubt, I took Lloyd for granted during our childhoods. At those ages he was kind of an assumed extension of me, two blood brothers with a common circulatory system. Cousin Jack, next door, who occupied the space of the year between us, was clearly a bonus, very much a free brother thrown into the package deal on our Knowles homestead. I always sensed that having five relatives next to us was something out of the ordinary, uncommon, fortunate. I also sensed and appreciated, even at an early age, the special roles played by Mom and Dad as they struggled against long odds to tuck their family into the folds of an ambitious dream. Mom and Dad had built a dollhouse against the storms of a demonstrably dangerous world, and I could sense the critical importance of their roles in bracing up the walls against the howling winds outside.

Only later, did I come to appreciate how much Lloyd was doing the same thing for me. He was there with me all the time, like a good Seneca guide, a few steps in front, first to peek over the hilltops and around dangerous looking rocks, then quickly calling descriptions or other important pieces of information back to me so that I would be a little more prepared for what was inevitably coming in our world. We were close enough that we shared most of the same experiences and tucked them away in the same parts of our hearts. Years later when we wanted to call them back to mind—the berry-picking and the football games, the Christmas rituals and the grapevines swung out over the ravines—we only had to say a word or two to put us both back in exactly the same place. Then we would nod knowingly, and smile. Nothing else needed be said.

~~~

The Cuyahoga River was not the only Indian highway in the land of my youth. As long as people have been in the valley there have been trails into and around it, thin slits through the monstrous, malarial jungles that took the Indians where the waters would not. The trails were not built for speed. The war paths, in particular, often went out of their ways to advantage themselves of open views so as to reduce the possibility of ambush. The by-product of this bit of pragmatism was a frequently scenic beauty associated with the trails. At other times, the trails would swerve awkwardly for no apparent reason. Behind the swerves might be the ghost of an ancient, fallen tree, long since rotted away, or access to a lake that dried up before later travelers were born. Nevertheless, the Indians continued to observe the swerves and detours as obediently as had their great ancestors who had good reason to do so.

In 1933, Frank Wilcox made his name synonymous with Ohio Indian trails as he tracked through backyards and over paved roads to recreate, at least in his mind's eye, the lines of the great old paths. His *Ohio Indian Trails* is a moody yet scholarly walk down the trails, journeys that saw him plodding past skyscrapers and grain silos to catch just the glimpse of a bronzed shoulder or a flapping loin cloth disappearing around a far tree. His mood is infectious. There is something magnetic about an old trail. Rivers and streams are always moving on, sweeping away pieces of history that touch their ancient waters. But trail lines and markers stay forever. They may be buried beneath heavy layers of the forest's annual composts, or plowed into neat rows of corn, or paved under boring stretches of highway, but they are still *there*. Sometimes the signs are more tangible, like the ancient candelabrum-shaped Signal Tree in Akron by the Cuyahoga Falls border that some believe marked the Indian trail to the Portage Path. But even without such evidence, you can sense the ancient trails if you're really looking for the Old Ones.

Fair warning: the next few pages of this work get tedious in the tracing of the old Indian trails. I bored myself rereading it (though the original research and drafting were exhilarating). Why is it so important to establish the probability/plausibility/possibility that one of the trails ran across the back of our actual property in Northfield, that our small feet trampled the same ground as our mysterious geographical ancestors whose echoes continue to haunt the nation? After all, couldn't the same claim be made of those in downtown Cleveland who walk Euclid Avenue every day where a trail also ran? No. For starters, there are yet places in the Cuyahoga Valley where an Indian could still feel at home, at least for a short while, the ravine being one of them. Not so for Cleveland. Euclid Avenue residents and businesses, like all city creatures, began by burying the old trails. Then they forgot about them. Now, most Clevelanders don't even suspect their former existence. But if these old trails contain no more importance than they do visibility, then this entire book is in vain. The claim that "we are shaped by our past" is uncontested but also ignored in our contemporary world. I guess most people simply don't take the time to care about what shape they are in.

For better or worse, I have to know—or know more.

The best known and most easily guessed Indian trail is the Lake Trail that follows Lake Erie's southern shore along the favorable rise of ground sculpted by an earlier level of the lake. It frequently follows the route now claimed by U.S. 20, and then Euclid Avenue into downtown Cleveland before angling sharply across the face of the Superior Street hill to the mouth of the river. The Great Lakes themselves were natural routes of transportation for the Indians. During the warm seasons a single Indian might paddle his canoe from one of the western lakes all the way to Montreal, an incredible accomplishment to anyone who has tried to negotiate even an air mattress on Erie's shallow, choppy waters. But, then as now, the weather wasn't always warm in northern Ohio, and the lake shores were anything but the shortest distance between two points.

There were many who preferred the shortcuts over land. Upwards of 2,000 Iroquois filed quietly along at least the upper parts of the Lake Trail enroute to their bloody rendezvous with the Eries. A century later, the pro-French Pontiac, perhaps coming eastward along the route, met a westward bound British Major Robert Rogers, who had the unenviable task of telling the French fort commanders that their war with England was over and lost. Some, including nineteenth century historian, Francis Parkman, believe that the tense meeting took place at the debris-choked mouth of the Cuyahoga River as the days shortened in 1760.[14] Within three more years, Pontiac's runners would be carrying war belts of wampum up the same Lake Trail as he tried to woo the fading power of the Six Nations (the Tuscarora joined the League at a later date) for his great uprising in 1763.

But the Lake Trail stayed pretty much north of us, and the point at which it crossed the Cuyahoga Valley is the point at which the valley ceases to be a valley at all—the short distance between Cleveland's Flats and the mouth of the river. That trail's significance was tied almost exclusively to the strange contours of Lake Erie and the other Great Lakes, the rivers that flowed into them, and strategic points among them such as those that would become Detroit, Chicago, and Ft. Wayne. It was a trail not easily sidetracked into the southern interior.

The story was similar in our immediate vicinity. The Portage Path at Akron was probably millennia old even before it began appearing on the earliest European maps more than 400 years ago.[15] But although many miles south of Lake Erie, the Portage Path owed its existence to the waterways of the Western Reserve. The eight-mile trek over the state's watershed connected the lowest loop of the Cuyahoga and the highest reaches of the Tuscarawas, the only dry spot along an otherwise all-water route between the Atlantic Ocean and the Gulf of Mexico (of course, most canoeists preferred to portage around the rather inconvenient cataract at Niagara). At times, the Summit County portage seems to have been as short as one mile. William Barnholth convincingly hypothesizes that an earlier and fuller Summit Lake at Akron reached much further north before the canal and other man-made changes turned the lake's northern shore into a trickling creek (Willow Run). Particularly during the rainy spring seasons of those earlier days, Barnholth argues, the old and baffling references to a one-mile portage become plausible, even probable.[16]

Wilcox identified at least three major Indian trails that stabbed into and through our part of the Cuyahoga Valley. From the south came the Cuyahoga War Trail and Muskingum Trail, that began their respective routes in the vicinities of Delaware (Ohio) and Marietta, joined together on the Portage Path, then ran up the valley along the western ridge of the river, probably close to present Route 21. Even closer to home was the Mahoning Trail. That well-traveled path came westward from the confluence of rivers that formed the Ohio (i.e., at modern Pittsburg), took care to advantage itself of the valuable salt lick near present Warren, then followed the northern Ohio watershed toward Akron. A fourth, smaller trail came our way from Aurora to our east after slicing down the line of the Chagrin River from the Lake Trail to the north and east.

Of these trails the last two are of greatest interest in terms of Northfield, in general, and our little nook in the southwest corner of that original township in the Western Reserve (now Sagamore Hills). The Mahoning Trail divided at Cuyahoga Falls, sending a ninety-degree spoke up the valley through Silver Lake, Northampton, Boston and Northfield before finally finding and crossing the river near the historic place where it intersects Tinker's Creek. From there it ran north to join the Lake Trail. The other branch of the trail moved due west to become the aptly named Watershed Trail that pushed on toward Upper Sandusky and other trail connections in northwestern Ohio. The northward thrusting Mahoning Trail passed by the head of Brandywine Falls and stayed west of town (Northfield Center) while making its way to the well known confluence of Indian trails at Willow Lake in what is now the Eaton Estates.[17] Wilcox believed that the trail followed the line of Rt. 8, but to do so it would have had to angle sharply eastward from the falls (passing through what is now All Saints Cemetery) for no apparent, geographical reason. A straight line, while certainly not a consistent characteristic of Indian trails, would see the trail running somewhere between Brandywine and Boyden Roads, probably close to the line eventually claimed by the New York Central Railroad, which is now a bike trail. This opens the fascinating possibility that a major foot road between the strategic sites of Ft. Pitt and Ft. Detroit ran within a thousand feet of what was to become my bedroom for thirteen years.

The Chagrin River Trail also offers an interesting possibility or two. From Aurora "a branch may have continued down Brandywine Creek to Ponty's Camp, one-half mile north of Boston Mills on the Cuyahoga."[18] Wilcox suggested that the other branch followed the line of present Rt. 82 through Macedonia and Northfield before angling north to the Willow Lake watering hole. Given the great tracker's uncertainty and this writer's wishful thinking, it is not unreasonable to think that the southern branch of this trail may have found its way, quite literally, through what would much later become our backyard.[19] Both the size of Brandywine Falls and the direction of the southward seeking trail dictate that it would have passed to the north of the falls on its way to Ponty's Camp, which is usually placed close to where Brandywine Creek empties into the Cuyahoga, and always on the west side of the river.

The ravine that runs parallel to that of Brandywine Creek, one ridge over, and hugging Highland Road, is the valley of my childhood. Along it lie Walls's Pond, Jensik's Creek, and the falls. It seems equally probable to me that the trail to Ponty's Camp, or a subsidiary, might have run along this ravine ledge. It would have angled down to join the Cuyahoga just above Brandywine Creek. Furthermore, it would have been the more direct route if the branching of the trails started not at Aurora, but rather farther west, perhaps at Northfield Center, an assumption based on the common practice of joining two trails together for any length of ground that could serve both trail purposes. Since the Rt. 82 trail seems logically older and more established than that which descended to Ponty's Camp (since the Indians' use of Willow Lake easily predates the Ottawas' encampment), it makes sense that the younger branch would make use of the older for as long as possible before breaking away on its southwesterly slant. Such an angle would take the trail branch across current Brandywine Road to intersect with the northward running Mahoning Trail at some likely point. Such a pragmatic point could well have been the natural spring a few feet east of the bike trail that initiated the flow of waters down "our" ravine to the Cuyahoga. (If the spring was active in earlier centuries it would also argue for the more westerly course of the Mahoning Trail that I have had the audacity to suggest as a slight correction to Wilcox's route through

Northfield.) From this point, the Indians would have had an easy hike down the ravine ridge to the Cuyahoga floor and Pontiac's Camp, a route that would have run them through the back corner of the property Dad and Mom owned for fifty-seven years. But there is another reason why I think an Indian trail ran along the ravine.

The path was *still there* in our time.

The path never made much sense to me. The only extant piece of it ran from the top of the falls for maybe fifty to one-hundred feet along the crest. But there is no reason for its being there, unless it is the faint outline of a trail pounded out over the course of centuries. (The ideal location of Ponty's Camp at the conjunction of a stream and the river made the camp a likely Indian habitation and, hence, trail destination, even in Erie times.) From the moment we arrived in Northfield in 1953 to Mom and Dad's exit in 2008, there was no regular foot traffic along the path. It was not even the regular entry to the ravine and falls for us kids since we usually made that entry via the ball field. Nor could it have been a likely route when the property was part of a farm in the distant decades before our arrival. Running atop the rocky shelf of sandstone outcroppings, the path could never have been particularly useful to either the farmer or his cattle. Shorn of Indian logic, it is a path that begins nowhere and leads nowhere.

∼∼∼

So, again comes the question: why is it so important to know whether an old Indian trail may have come within a hundred feet, rather than a thousand feet, of our house? And, for that matter, why is it important to me now? To be truthful, very few of our childhood games had to do with the ancient shadows of Indians still lurking in our woods. Too, my fascination with the Indians notwithstanding, we didn't spend much time thinking about them when we were kids.

I suppose a part of the answer lies with my unbounded fascination with trails and paths of any kind. Any swath of tamped weeds or any corridor through coincidentally spaced trees instantly fires my imagination regarding the identity of the trailblazers. Human paths imply human purpose, but there was little other evidence of such purpose in most of the 2,000 acres of wild woods surrounding my youth. Along the rare trails lay the endings to stories that I wanted to know. One such story involved a semi-mythical "Stagecoach Road" that teased us with fleeting hints of its existence. Lloyd, Jack, and I talked about it as a reality, feeding our conviction every time we stumbled upon an unexplainable tire rut or an odd opening among the trees. Tuck Richards, a boy who lived in the big house at the top of Brandywine Falls (now a well-known bed-and-breakfast), showed us further proof in the forms of wheel marks in the woods and a pile of wood that he expertly identified as the remains of an old stagecoach. We never thought to ask ourselves why a stage line would run between two endpoints that ended in nothingness, just as I never considered the probability that a stretch of tamped weeds had resulted from the frisky romp of a wet dog fresh out of the creek. Then, several decades later, my Mom solved the delicious mystery when she sent me a *Cleveland Plain Dealer* article about the old Marshall Estate that included a detailed description of an old carriage road that the family maintained through parts of the vast estate. As I said, "semi-mythical."

But a bigger part of the answer, I suppose, is a larger and, hopefully, more mature fascination with the pieces of lives that people leave behind them. Generally speaking, the fewer the pieces, the greater the allure.

The canallers left an old lock, a stretch of extant slack water, and now have a museum in the national park to look out for their interests. Our railroaders adapted to change by offering their long, straight beds as a bike trail and tourist excursion line, and the parent transportation mode still does a good business across the land. But the leavings of the Indians require an expert eye to see the arrowhead buried among a pile of small stones, or the faint outline that may have been part of a trail noted on a map centuries old, or the ordinary looking hill that contains the bodies of hundreds of Hopewells.

Ah, the Hopewells. They are the only Indian nation that most people associate with some facet of actual Indian culture—mound building. Of all of the mound builders, the Hopewells were the most interesting. Although preceded in Ohio by at least four earlier Indian types (Paleo, Archaic, Glacial Kame, and Adena[20]) who warily filtered in on the heels of the receding Wisconsin Glacier, we know the most about the Hopewells, whose Ohio tenure roughly spanned the millennium between Aristotle and Charlemagne. Ironically, what we know about Hopewell life came from their infatuation with death. Bloetscher says that the Hopewells:

> became so obsessed with the subject of death that they organized a 'Cult of the Dead,' wherein the energies of the living were spent largely in funeral preparations for the deceased. Most of the building was for the dead, much of their art was created for the dead and a great network of trade routes was established, to provide raw materials used in making objects for the dead.[21]

It would be easy to criticize the foolishness of a people who spent their lives in preoccupation with death, but I wonder what the archeologists two millennia from now will find of the preoccupations from our lives. At any rate, even the Hopewell death leavings are sketchy in the Cuyahoga Valley. While southern parts of Ohio boast of abundant evidence of Hopewell presence, the valley evidence is far less convincing. Bloetscher parenthetically notes that hearsay has traditionally ascribed the Star Mound near Riverview Road to Hopewellian origin, and that the Gleeson Mound farther down holds similar promise. Jesensky states that Egypt Mound, across Dunham Road from the Walton Hills Church of Christ, hints of Indian origin, and that Indian Hill a few hundred yards upriver from Red Lock Hill has been oozing Indian artifacts for the better part of two centuries. More recently, Cuyahoga Valley National Park archaeologists have also unearthed evidences of Hopewell homes in Everett Village.[22]

If the physical leavings of the Indians are scant, so too are indigenous records of their character. They left remarkably few etchings or inscriptions, the possibly-Erie petroglyphs preserved in stone on Kelley's Island in Lake Erie being a rare exception. Eckert repeatedly affirms that the detailed wampum belts circulated among the various tribes could be read with great precision, often providing the texts for lengthy and complicated speeches at Indian councils. But these never made much sense to whites. Our Western definition of history—that it begins with the recording and keeping of written records—means that most of the North American natives were "prehistoric," and that even the well-known Shawnees, Ottawas and Iroquois would have remained such had they not come across white Jesuits and other chroniclers who could handle the clerical chores on their behalf.

A good part of the tragedy concerning the American Indians is that so little of their moral tenor was passed along for succeeding generations, red or white. In just our small corner of the Western Reserve, we are enormously indebted to British and Blacks, Germans and Jews, and a dozen other ethnic groups that contributed to a mix of life not quite the same anywhere else on the planet. But when we drove the Indians off we threw out the baby with the bath. We could have learned something, for example, from the Indians' sense of sexual fidelity, and more importantly, the reason for it. The husband took great pride in being a good provider for his wife, and she was equally determined to show her love for him through hard work and tedious service. The stunningly simple result of this selflessness was that "divorce was naturally rare, since both partners worked so hard to please one another."[23] I wonder if any marriage counselor has ever improved upon this concept. Regarding racism, there is little evidence that the Indians initially resented the presence of the whites, and in fact, usually treated them better than they did neighboring tribes with whom they maintained bickering feuds for endless decades. Furthermore, they probably never really felt a fully justifiable sense of outrage over stolen or swindled lands because they did not see ownership in the same way as the whites did. The Indians grew to hate the whites not so much because of the latter's being there (most of the tribes got along quite well with white French traders who adopted many of their ways) but because of what they did there. It is a story that might have had a better ending.

But, of course, it didn't. And so we wonder about the might-have-beens, about opportunities lost. In their fading historical wakes, we hear the echoes of their spirit, but only when we bother to listen.

Notes

1. Virginia Chase Bloetscher, *Indians of the Cuyahoga Valley and Vicinity* (Akron: St. Mary's Church, Anglican Catholic, 1987), 111. There is some inevitable apples-to-oranges measurement difficulty here. Over the years the U.S. Census has made changes in the way racial demographics are reported. Bloetscher's figure probably includes Census counts of mixed race categories. For 2013, the Bureau's State & County Quick Facts listed approximately 1,100 Indians for the entire county, but that figure was limited to "persons reporting only one race," and included Alaskans.

2. Thomas A. Bailey, *The American Pageant: A History of the Republic* (Boston: D.C. Heath and Company, 1966), 44.

3. Harry Forrest Lupold, *The Forgotten People: The Woodland Erie* (Hicksville, New York: Exposition Press, 1975), 55.

4. Francis Parkman, *France and England in North America:* Volume 1 (New York: Literary Classics of the United States, Inc., The Library of America), 1983.

5. Ellis, *The Cuyahoga*, 18.

6. Lupold, *The Forgotten People*, 46-47.

7. Ellis, *The Cuyahoga*, 88.

8. Allan W. Eckert, *The Conquerors*, (Boston: Little, Brown and Co., 1970), 305.

9. Eckert, *The Conquerors*, 190.

10. Eckert, *The Conquerors*, 82.

11. Allan W. Eckert, *The Frontiersman* (Canada and the U.S.: Little, Brown, & Company, 1967), 444.

12. Eckert, *The Frontiersman*, 539-540.

13. Bloetscher, *Indians*, 22.

14. Frank N. Wilcox, *Ohio Indian Trails* (Cleveland: The Gate Press, 1933), 64.

15. Ellis, *The Cuyahoga*, 3.

16. William I. Barnholth "George Croghan, Cuyahoga Valley Indian Trader" (Northampton, Ohio: the Northampton Historical Society, undated), 15.

17. Wilcox, *Trails*, 72.

18. Wilcox, *Trails*, 120.

19. It is worth noting that much of what we know about the trails is based on guesswork, and that the experts may disagree about exact routes. For example, Joseph Jesensky, in his "Archaeological Survey of the Cuyahoga River Valley," plotted the Sagamore Trail—a most important trail long before the white man's earliest roads were made—as a hypotenuse-lined shortcut off of the Mahoning Trail slanting west of Northfield to rejoin the main trail near the juncture of Tinkers Creek and the Cuyahoga, a site placement that Wilcox appeared to affirm on his earlier maps without comment. But Bloetscher shortened the Sagamore Path so that it rejoined the Mahoning south of us at Boston Ledges.

20. Bloetscher, *Indians*, ii.

21. Bloetscher, *Indians*, 13.

22. Cuyahoga Valley National Park website, *American Indians*, information retrieved March 10, 2015, http://www.nps.gov/cuva/index.htm

23. Bloetscher, *Indians*, 81.

6
Life Down Under
The Living Things

Doe tending to her fawn. CVNP. It sounds surprising, but I do not remember ever seeing a deer during my childhood romps in the valley. Years later, after Lloyd and I were long gone from home, groups of them would pay daily visits to Mom and Dad's place, sometimes resting under the pine tree in the front yard. I think the reason is simple: in our time, the huge stretches of woods and ravines—private property, then—were largely ignored by all but a few wandering kids, allowing the animals far better options for seclusion and forage. *Photo by Tom Jones.*

The "critter" history of the early valley • The "Great Hunt" • Of dogs that never die

One unremarkable day, circa 1960, Lloyd, Jack, and I were fishing in the eastern reaches of Brandywine Creek, close to where the Akron-Cleveland Road (Old Route 8) and the historic creek intersect to quietly mark the forgotten community of Little York, one of the earliest settlements in Northfield Township. There was nothing left of Little York on that day more than fifty years ago—I had never heard the name until I began researching the township decades later. This beautifully secluded short stretch of the creek was also marked for extinction. Indian Creek, which fed the Brandywine at this point (and which the earlier folks called Indian Run), was about to be chewed up by the Indian Creek housing development just a hundred yards to the north. Meanwhile, Brandywine Creek itself had already been blue-lined by the planners of Interstate 271 who would straight-jacket its sleepy curves into tons of concrete. But for that moment, on that day, in that one spot, Brandywine looked much as it had for the previous ten thousand years or so.

The stillness of the setting had lulled me into a trance. The fish weren't biting, and Lloyd and Jack had spread out to other locations out of my view. My red and white bobber did its ancient magic, appearing at once to be both flowing with the current and standing still. It was the kind of moment when it is hard to distinguish between seconds and minutes. The spell was suddenly broken by a loud thrashing off to my right. I turned just in time to see a small beaver flop noisily into the glassy waters and glide downstream. What I had seen was, of course, not possible. I was no more likely to see a beaver in these parts than I was to see a black bear or a panther or a wolf. I yelled for Lloyd and Jack, but when they came they did little to encourage my report. It couldn't have been a beaver, they said. It must have been something else. But I could think of nothing else that size, including a dog, that went into the water with that kind of grace and confidence. I was absolutely *sure* it was a beaver! Wasn't it?

The many ensuing years have rekindled the possibility. The beaver have, in fact, returned to the valley. There are many choice spots for them in the Cuyahoga Valley National Park, and the extreme narrowness of the Cuyahoga River means that the park rangers have to keep close watch lest a pair of the industrious critters clog up the artery between Cleveland and Akron. My problem is that I am not sure anyone would credit a return as early as 1960. My guy would have to have been an especially eager beaver. But there were no park rangers to monitor the wildlife back then, only an occasional kid or two. And the thing sure looked like a beaver.

No matter. On the zoologist's clock, the difference of a decade is but a moment. Furthermore, beavers are not natural strangers to the valley. Until about 1700 they were as much a part of the Cuyahoga landscape as deer and rabbits. Economic historians claim that these toothy residents were largely responsible for the great wars between the Eries and Iroquois in the mid-seventeenth century, as well as the wars that the Iroquois waged against the Hurons, Neutrals, and Tobacco Nation in Canada. When the beaver grounds in the Iroquois' native New York played out under enormous profit pressure from the British-based colonies in the east, the Six Nations attacked the western interior driven by the same forces that have driven hundreds of wars among the whites. They would either conquer or fall into ruin; there was no standing still.[1] The route of that conquest would necessarily lead the Iroquois through the beaver-rich Cuyahoga Valley.

The extermination of the Cuyahoga Valley beaver notwithstanding, the Indians usually lived comfortably with the other living things sharing their lands. Whether plant or animal, the natives took only what they needed, and treated special hunting grounds with reverential care, including a great hesitancy to live on them. The huge Kan-tuck-ee land south of the Ohio River was the best example of this kind of Indian hunting reserve, which explains why they were so enraged when the whites blatantly encroached upon it. The Cuyahoga Valley may have been such a hunting ground, although this residential timidity probably had as much to do with fear of the Iroquois absentee landlords than with any sense of the sacred. But the hunting was very good. With the felling of a tree or the killing of an animal, the Indian was likely to apologize to the life he had just taken, often remarking that he would not have made the kill had it not been absolutely necessary. The Woodland Indian then proved his sincerity by making use of almost every part of the fallen creature. The deer's skin provided moccasins and clothing, its hair ornaments and embroidery, the antlers tools and arrow points, and hooves made glue and rattles. Tendons and other sinews became thread and bowstrings, the bladder served as a bag, and the animal's bones were used to dress the very skins that once hosted them.[2]

The white pioneers in the Western Reserve were less appreciative of their new, natural neighbors. The bears, panthers, and wolves, which, like the beaver, exited our land long before the Knowleses' arrival, were still present in great numbers when the first settlers came in the early nineteenth century. Pigs and chickens in those pre-fencing days did not have much of a chance when the poor farmers, sometimes facing starvation themselves, were forced to turn the animals loose to forage. Neither were the predators shy about inquiring at the house when the farmer's livestock was sparse. A full butter churn or a piece of hanging venison was always appreciated. But it was not a completely ill wind that blew in from the wilds. The predators themselves became attractive prey, especially the bears. Spencer Phelps, settling in what is now neighboring Geauga County, stated that he forced nine bears to make the ultimate sacrifice for their new-found taste for pork and chicken. In general, the pickings must have been pretty easy all the way around. The wolves were so unintimidated by the first white settlers that their eyes could be seen shining through the cracks of the log cabins at night. Deer were sometimes caught by hand.[3]

Rattlesnakes did not even qualify as a mixed blessing in the settlers' minds. By all accounts, the early Western Reserve was swarming with them. Attracted by the warmth of a hearth and undeterred by porous cabin walls, the rattlers often joined families when they curled up for a night's sleep. Sometimes they simply and suddenly reared their heads up through cracks in the primitive flooring. Men routinely carried staffs, ready at the first sound of the telltale rattles to cudgel any dark-skinned sticks not attached to a tree. Stories are handed down of men who killed so many of the snakes in one spot that, even though unbitten, they took to their sick beds as a result of the poisonous miasma in which the battle had been joined. Someone bothered to record what might or might not have been the record for rattlesnake kills: 485 on a rocky ledge in Trumbull County.[4]

Very soon, they were all gone—the bears, wolves, panthers, and most of the rattlers. In all my years of romping through the vast wooded tract bordering our property, I never saw a rattler, nor anything more serious than a big black snake sunning himself in an shaft of light penetrating the forest. Uncle Frank thought he had killed a rattler once. After he halved it with a garden spade he showed us the continuing vibration of the tail tip, but I couldn't tell if that was a rattle or just the creature's physical reaction to a violent death. At any rate it was far short of the 24 rattles that Abraham Norris counted on the rattler he killed two centuries ago. Norris slew the beast with an ox goad, inside his cabin, moments after his wife took five years off of his life with a scream that a banshee would have envied.[5]

There is far less doubt about the removal of the large predators. I first suspected as much one day in the third grade when, in a fit of childish one-upmanship, I decided to flavor my annual "What-I-Did-on-My-Summer-Vacation" oral report with the comment that I had been chased by a bear. Once I had gotten my teeth into the lie I could not let it go, even under the incredulous stares of my classmates. Mrs. Oviatt, having taken the college education block during a day when "No Nonsense 101" was a staple of the curriculum, rather firmly convinced me otherwise. Yet, I wondered why I had not, in fact, ever seen a bear in those woods. It looked like the kind of place where a bear would be. But since bears and bobcats were among the few things Mom and Dad had not warned us about in this life, I had to assume that Mrs. Oviatt was on good grounds when she backed me down. The bears were gone, as were the wolves, and the big cats. The woods was big, but not big enough for both those kinds of hunters and the most dangerous predator of all.

There were other animals that did learn to live in our uncomfortable shadow. In spite of the fact that our couple of acres was dead center between two of the nation's larger metropolitan areas, cupping two and one-half million people within a half-hour's drive, our place still teemed with wildlife. Decades later, Dad was still being frustrated by the deer who sauntered up casually to his flowers and bushes for evening snacks (he called them the "goats"), driven to greater boldness by the many real estate incursions into their formerly undisturbed lairs. The area is a riot of rabbits, groundhogs, opossum, chipmunks, and skunks that made the necessary adjustments where the bears and wolves could not.

There is, of course, a third part to the animal story of the valley, and that involves the domesticated variety. Arguably, Ohio's history would have been radically different had it not been for horses, oxen, and mules. Without them, Cleveland would probably be living in the shadow of Sandusky or Lorain, and there is a fair chance that Akron would be nothing at all. The beasts of burden quite literally pulled northeastern Ohio into a position of prominence, both regionally and nationally. They hauled the monstrous stumps that stood in would-be farmers' fields, and pulled the crude plows that gouged the forest floors to channel the Ohio & Erie Canal and its eighty-ton freighters that slid through the great ditch in what would become the mightiest commercial enterprise in the world's history. When Isaac Bacon, Northfield Township's first householder, set up housekeeping somewhere near where the Eaton Estates now grace Valley View Road, he had no one other than his oxen to haul the logs into place as the walls of his cabin climbed beyond the limits of his reach or strength.[6] Other animals not suited to pulling made their contributions by feeding, warning, and protecting the families living on the razor's edge of existence. Sometimes they became valuable currency in the give-and-take of the primitive valley economy, to be bartered for a piece of land, a cabin, or a plow. From a purely practical viewpoint, a family could suffer the loss of a small child or aging relative easier than it could that of their mule.

~~~

Hinckley is a small dot on the map that lies along Rt. 303 a couple of miles west of Richfield. Other than being the place where a jagged piece of a Coke bottle gave me a lifelong toe scar at Hinckley Lake, the town is notable for only one event: the return of the buzzards every year on the Ides of March. The local folks make a bit of a to-do of the occasion with buzzard watches and pancake breakfasts, while wildlife officials speculate on when or if there will be any sightings on the appointed day. Apparently Hinckley residents aren't the only ones caught up in the excitement as the town population swells to some thirty thousand people for the big day. Across the state, journalists treat the event as an article of faith for their March 15 editions—you can probably find at least a paragraph in most of Ohio's newspapers—but the less excitable types simply shrug and note that the buzzards, like a lot of other migratory birds, just have a way of coming back about the same time each year.

Certainly there are more spectacular evidences of nature's cycles, or at least more romantic ones. The swallows that annually grace Capistrano draw the hearts of lovers and the souls of poets. Buzzards draw flies. To be sure, there is nothing for a town to be ashamed of in identifying with the turkey buzzard. The bird is a hard worker and serves an entirely necessary purpose. In these and other ways, he may be a good symbol for the reserve. But it is doubtful that Hinckley's infatuation with the buzzards could have sustained itself on the back of such simple pragmatism. In fact, there probably would be no Buzzard Day at all in Hinckley had it not been for the story many people prefer as an explanation for the annual punctuality of the great hawks. According to the story, the buzzards return every year because somewhere deep in the recesses of their DNA is the corporate memory of a feast such as no buzzard dared to dream, before or since. They return to Hinckley, site of the original and delicious carnage, on the odd chance that something like it might happen again. So say the old timers of Hinckley.

The "Great Hunt," sometimes called the "Grand Hunt," was, itself, no legend. It took place, not on March 15, but rather on Christmas Eve in 1818, not too many years after the first whites began filtering into that part of Medina County. Even then, Hinckley had a reputation as a "hunter's paradise" among settlers, and a few years earlier, the Wyandots and Senecas as well.[7] But there was a problem: the local bears and wolves were using the place as a base from which to operate against the livestock on nearby farms. The solution was a military style campaign that exceeded anything seen in those parts during the three great wars waged among and between whites and Indians during the previous sixty years. Nearly five hundred men reported for duty in what was, from the beginning, more of a hoot than a fight. Older accounts of the event dwell on the social successes as well as the practical outcomes of the Grand Hunt. It was surely the largest block party in the history of the Western Reserve.

The general idea was to form four walls of hunters in a rectangle covering about a half square mile around the center of Hinckley. There is no mention of Northfield hunters participating in the Grand Hunt, but it is likely that some did since Brecksville was one of the major contributors to the eastern line of the hunt. Bath sent men in on the southern line of attack. From the north came hunters from Cleveland, Royalton, and Newburg, while Strongsville, Brunswick, and Medina saw to the western line. Captains, men of note such as Judge Welton of Richfield, were appointed to keep the assaulting columns in a rough semblance of order. At the sound of a horn echoing down the valley, the men moved forward. The obvious question is which of the animals, four-legged or two-legged, were put in greater peril by the attack. Had the barrel of whiskey that graced the later victory celebration arrived a few hours earlier the thing might have ended in a bloody draw. The sight must have looked like something out of a federal OSHA staffer's fantasy nightmare, slaughtering every safety standard known to common sense.

> Steadily the columns press on, silently at first, then comes a wild shout and soon the echoing roll of musketry, as the wild game dashes through the woods and the thick underbrush before the advancing host. The north column is the first to close in the square on the center, then follows east and west and south. It was almost a solid phalanx of men, standing close to one another.[8]

No one pushed his luck farther that day than William Coggswell of nearby Bath, acknowledged as the "prince of hunters." Coggswell was ordered to take a few men and advance to the center of the lethal square in order to drive the half-crazed wolves and bears into the lines of fire. This he did most effectively until he got sidetracked by the appearance of "a monstrous bear—I think the largest I ever saw of that species." The prince of hunters was not likely to resist such a prize. Unfortunately, neither were the peasants hunting with him. Coggswell recalled:

> We met the bear just as he was crossing a little creek on the ice. I ran up the bank within twenty-five or thirty feet of the bear, and stood several feet above him. About this time the men on the south line commenced shooting at the bear, apparently regardless of me and my dog. There were probably 100 guns fired within a very short space of time, and the bullets sounded to me very much like a hail storm.[9]

Amazingly, only one man was shot during the lead storm, and it was not Coggswell. It is hard to say what factor rendered this good fortune to the oblivious hunters, but I am less than convinced, as was one historian, that is was the strict order to the men to fire "low and toward the center, to prevent injury to the men." "Low," in fact, is one of the first pieces of firing advice military leaders give their underlings when instructing them on how to shoot other men. Unless the Hinckley huntsmen possessed some unique capacity for hovering five or six feet off the ground, "low" was an open invitation for friendly fire once the constricting walls came close. And the idea that firing toward the center of an enclosed, human rectangle somehow increased safety does not even merit a response. More likely, the hunters could thank the animals for the low casualty rate. The final death count was more than three hundred deer, twenty-one bears, and seventeen wolves. No census was made of the lesser critters shot that day. The tightly packed mass of animals must have served as a protective shield for the men, each of whom was, by design, blazing away in the direction of at least one of the other hunters.

But if the hunters' luck was good, it was nothing compared to that of the carrion creatures. Those that were patient enough to wait for a cessation of firing and a commencement of the drinking must have gotten a good jump on their Christmas dinners. A local bard from Cleveland would have us believe that the hooch dragged in for the celebration had no negative effect on the men:

> "Then every man drank what he chose,
> And all were men of spunk;
> But not a fighting wrangle rose,
> And not a man got drunk."[10]

Uh-huh.

Medina County purists might believe that, but I doubt the surviving animals in Hinckley's dense underbrush did. It is well documented that the big bear got roasted and consumed during the party, but there is no record of what was left behind when the cold, exhausted, hunters gathered up their weapons and their headaches to head for home that Christmas morning. The leavings must have been considerable. The problem with the story here—perhaps this is why reporters insist on using the word "legend," in describing it—is that the migratory turkey buzzard would, on this occasion, have been preparing for his Christmas in Gatlinburg, Tennessee, not Hinckley, Ohio. But, who knows? Perhaps there was one old bird that had stayed behind because he knew he could not make the long flight south. Or maybe a long, cold winter preserved the feast until March 15 when the returning flock—not especially noted for their intelligence—assumed that the glorious killing had happened on that very day. Neither explanation stretches credibility to the breaking point. Believing that story is only a little more difficult than believing that the buzzards' return to Hinckley every March 15th would turn out thirty thousand people to look for them.

~~~

Hunting was never a part of the Knowles family tradition. From time to time, Uncle Frank would tack up "No Hunting" signs, and we listened tensely as the shotgun blasts ripped the air in the vast woods around us during hunting season. One time Jack and I found a pile of spent shell casings in the entrance to the woods near the Manseys' house up the road. We were primary-schoolers at the time and assumed that we had stumbled onto something that would unlock the secret of the always-heard but never-seen hunters. I am sure that safety considerations had much to do with our anti-hunting feelings (we got the safety lecture from our parents every year when hunting season rolled around), but that was only the formal reason. Far more important was our sense that hunting violated our belief that nature is a landlord, not a realtor. Our delightful duty was to live in harmony with God's endlessly fascinating creation, not to subdue, manipulate or, heaven forbid, destroy it. Except where our ignorance made us clumsy, we took up only so much room as we needed. The other creatures needed room, too.

This line of thinking is getting more attention now, like something you might hear a celebrity say during a concert, or a politician during a speech to an animal rights group. But it was not much in vogue during the 1950s. That was yet a world of seemingly unlimited resources, where contractors used only thin wisps of insulation in stud walls because heating fuels were cheap, where the idea of recycling would have sounded horribly primitive (not to mention unsanitary), and where we assumed that we knew just about all there was to know about animals. They were just part of the limitless resources at our casual disposal. All fear of annihilation was focused on The Bomb; most of us never dreamed that we were risking the catastrophe much more openly in our bathrooms and kitchens.

But we were neither animal rights militants nor environmentalists before our time. And we would have had some difficulty answering the questions from the farmer whose corn was being vandalized by an exploding deer population. The truth was that we simply held wildlife in high esteem, and even a bit of awe. That is why Dad could muster only half-hearted efforts to protect his trees and plants from the marauding deer that casually dropped by each night as if they had dinner reservations. It is why, in the darkness of a long ago night on Boyden Road, I insisted that we back up the car two or three times to make sure that we had killed the suffering opossum whose nocturnal wanderings had gotten him tangled up with our car tires. It is why I took such delight in getting a baby skunk to take bread out of my hand after its mother had apparently met with some misfortune, forcing the tiny siblings to take to our ball field in search of food each evening.

Somehow, the rules of the land seemed to hint that these creatures had as much right to be there as we did.

~~~

What's in a pet? In an age when you can have your dog serviced by a manicurist, treated by a psychologist, and buried in a lead-lined casket, the question is no longer as simple as it once seemed. In frontier times, the answers were easy enough. Cats were prized for their ability to control threatening rodent populations, dogs for their alertness as guards. Other animals not typically thought of as pets all served very pragmatic purposes by doing, warning, or dying. The thin margin of existence left little room for keeping animals for sentimental reasons. But by the 1950s, such reasoning was a distant memory. After the stark horrors of the Depression and in the boom generated by World War II, Americans were no longer in a pragmatic mood. An entirely new economy, fueled by advertising and based on the assumption that people could be persuaded to buy more than they needed and most of what they wanted, guaranteed changed roles for the cat on America's window sill and the dog curled up on a corner rug.

But pets also faced uncertainties in the folds of our consumption-obsessed society. Values are debased and life is cheapened in a world where everything is viewed as a thing. Throwaway cameras and flashlights mean throwaway pets. If wives and children could be ditched when they became wearisome what better hope could there be for animals? I distinctly remember my sense of sickening shock when I first heard about people putting their live pets in paper bags and dropping them off on lonely country roads.

I don't know what prompted Mom and Dad to keep a supply of pets in our lives. It would be tempting to conclude that they wanted us to learn about life and responsibility—and those lessons were learned in the process—but I suspect they were primarily influenced by the same two reasons that induced most parents of their generation to give pets to their kids, those being the joy of giving gifts to their children, especially gifts that they had not enjoyed in their own bleak childhoods, and availability—someone was always trying to get rid of a litter of less-than pedigree puppies or kittens. Too, there was something about a dog at the hearth that seemed to be a part of the dearly nurtured American Dream that followed the GIs home from Europe and the Pacific, even when there was no hearth. I don't remember many of my friends who didn't have some kind of animal in residence.

Dogs were the mainstay of our Knowles homestead. There was a cat or two along the way, but these never lasted very long. In those pre-kitty litter days cats spent most of their time outside, which was OK in the summer, but dangerous during the snowy winters when roving dogs could too easily catch them in open territory. But we would have chosen dogs anyway. Our cats were boarders, but the dogs were family. With one exception, I cannot remember how we parted company with any of our cats, but the way we lost each dog is seared into my memory. Inky started our procession of canines. On the way home from the dog pound, Janet named her Inky, more, as I recall, because the name sounded tiny and feeble than because of any coloration characteristic. Technically, of course, Inky belonged to Uncle Frank's family, but that was not critically important at the time. The dogs never did understand the nuances of property lines, and besides, it was the pet precedent that was important. Inky was no superstar, but she had started a procession.

Much more lively and lovable was Laddie, also an FSK (Frank Stanton Knowles) dog who was, I think, part Collie, though considerably smaller than the full breed. Laddie was a good dog, but in the tradition of all tragic heroes, possessed one fatal flaw: he liked to chase cars. (That pet curse followed Cousin Jack into his young adult life. His family's dog, Toby, played four-bullet Russian Roulette by dashing out after cars zooming by on their winding, Tennessee road. Toby was hit numerous times, including once when his severe injuries prompted Jack to proceed with the grave-digging chore, an action that induced Toby back to life but not to wisdom. A few months later, he lost the last of his lives.) I don't see dogs chasing cars much anymore, which I am sure some aging philosopher of society will interpret as a sign that the modern generation of dogs is becoming physically soft and morally bankrupt—well, maybe that and leash laws. Virtuous or not, the practice was one that cost Laddie dearly.

Laddie's number came up on the rather sharp turn on Glencrest Drive in front of Uncle Frank and Aunt Doris's house. A few years later, Laddie might have been spared at least the means by which he died, that being the truck of the Meyers milkman, who was then still in the practice of making home deliveries. The truck was a cream-colored, humpty-dumpty looking thing, not unlike the cartoon vehicles that are always a bit too high and too fat to be safe anywhere except in the imagination. (Once in a long while, we would see a similar truck come by, but it would be brown in color. Naturally, I was convinced that this other truck must carry chocolate milk.) The milkman was already on our suspicious persons list for several reasons. He usually took the turn much too fast, about which our parents would occasionally warn us, and once had managed to get himself stuck in a deep mud hole down the road by the Copans' house. Dad tried to pull

Connected: Lloyd, me, and cousin Jack. Linked by physical proximity, shared experiences, and the rich bonds of brotherhood, we were known on the homestead as "the three little kids." The bonds would last a lifetime. *Photo from author's collection.*

him out with our little '51 Ford, but only succeeded in burning his tires bald. The milkman did not offer any compensation for the noble effort. Surely this was a bad man.

I did not witness Laddie's fateful tangle with the speeding tires of the Meyers Milk truck. In fact, I was one step behind the action all during that long day. Someone said that David threw himself down on the dog and cried, and that one of the other kids supposedly had a few words for the driver. But I also remember hearing that the driver felt bad about what had happened, an image that elbowed for room with my prior and more comfortable picture of his villainy. Later that day, they buried Laddie, and it is this part of the tragedy that lingers in my mind. For some unremembered reason, I was not with the others when they left for this solemn ceremony.

"They're burying her back in the field," Aunt Doris told me.

By all logic I should have assumed that "the field" meant the ball field on the back part of their property. Yet, inexplicably, I never quite found my way back there, instead spending several minutes wandering around the edges of Uncle Frank's garden far away from anything that could be considered a field. I vividly recall my feelings of desperate frustration as I realized that I was missing out on something of vast importance, the kind of frustration I have revisited many times in dreams. For many years thereafter, as I would gaze over at Laddie's grave by first base during a ball game, I could not shake the feeling that I had been cheated out of an important memory.

One of our first LCK (Lorence Cook Knowles) family dogs was a magnificent, large collie named Ruffie. We had gotten him from some other family that must have been in some kind of transition since I remember the impression that his stay with us was, from the beginning, something of a probationary period. Eventually, Ruffie would fail the trial, but what a wondrous and mysterious dog he was in the interim. Indoors, Ruffie was a gentle and good pet. His only indoor weakness, and that one was temporary, was a rather lively interest in the cat that once prompted him to lift the entire sofa with the bridge of his inquiring nose as he sought a better inspection of the ancient prey. But after dark and outdoors, something of the werewolf came out in him. Two chow dogs that I had never before seen in our neighborhood appeared nightly to wait for Ruffie when Dad let him out for nature's final call of the day. The ensuing fights were awesome, though we could only go on what we were hearing. Inside, Lloyd and I shivered at the sounds of unrestrained fury and pure violence, our first hint, perhaps, that something sinister lay deep within all creatures—even gentle Ruffie—something repressed, pushed back toward the distant past that it once ruled with raw, unguarded force.

Eventually, Dad would somehow manage to drive the chows away and Ruffie, torn and bleeding, would prance back into the house as if the whole thing was just part of the job. I was a bit awed by that noble sight. Here was a creature that charged headlong at what should have been his worst fears. He was alone, at night, in strange territory, set upon by two powerful foes that might well have come from hell, itself, yet he fought. It was a lesson that made an early and lasting impression. Too soon we were making expensive trips to the vet's to have Ruffie's wounds treated, and Mom and Dad began wearing the pained looks of decision-makers. Ruffie was just too much dog for the house, they said. Besides, they had not volunteered to train prizefighters. So, Ruffie went back to his original owners within a few weeks of his coming. The chows wandered off into someone else's childhood memories.

Unlike Ruffie, Tippy made the final cut. She was a beautiful cocker spaniel and a gentle house pet. Mom liked her because she obediently kept to a blanket behind the front door in the living room. Unfortunately, Tippy was also cursed by the short tenure that followed all of our early dogs. The only particular incident I remember about her life was her death, or more accurately, the discovery of her death. She had been missing for a couple of days, a sure sign of trouble for a house dog. Uncle Frank finally found her body in the crawl space under his house. My first reaction, even before grief, was one of bemusement. As we had searched the neighborhood for her my overwrought mind had begun to spawn visions of the vastly distant places to which she might have wandered. But all along she had been lying peacefully just inches under the feet of Uncle Frank's family.

He dragged her out and we crowded around her lifeless body. Someone guessed that maybe she had been poisoned. Naturally, I suspected that the foul deed had been done by Aggie Bardow, our somewhat unhinged neighbor to the east, after which poor Tippy had crawled under the house in her death agony. If Aggie was crazy enough to tell everyone that we had killed her baby, she was crazy enough to kill Tippy. But Uncle Frank shook his head and pointed to a thin line of blood along the lower part of Tippy's stomach. That didn't look like poison, but neither did it look like a fatal wound. We stood around scratching our heads. No one could come up with a satisfying answer. I recall being slightly hurt that she, for whatever reason, had chosen to die under Uncle Frank's house rather than ours. This time I made sure I was there for the burying ceremonies, which I think took place just inside the wooded undergrowth that met our lawn in the back. For a couple of years there was a marker on Tippy's spot, but then it was brushed aside and forgotten, leaving behind a dim, golden image of a fluffy cocker spaniel whose death, like Laddie's burial, would forever remain a mystery for me.

Canine longevity was finally established on the homestead with the coming of two dogs to Uncle Frank's family named Taffy and Zippy. Taffy, named for her color, was what I would call an opossum terrier, even though I don't think there is such a thing, and even though Janet would probably not like that description. But if you had seen ratty Taffy you would know what I mean. She was Janet's dog, but that was about the only thing Taffy had going for her. She was forever tucking her tail between her legs and veering out of the way whenever anyone came within twenty feet of her. If you tried to touch her, Taffy's options were limited to snapping and quivering, neither of which conjured up warm images of man's best friend. Later in life, she developed a pretty bad case of mange, which only reinforced our resolve to stay away from her. But Taffy was a good pet for Janet, and I guess that's worth a mention.

Zippy, who came along about the same time, was more like our kind of dog. He was affable, relaxed, and entirely appreciative of the free, rural environment around him. The only problem with Zippy was that he was misnamed. If ever there was a creature that "could not keep it zipped," he was the one. His sexual behavior was, to say the least, unrestrained, and anything from a legitimately in-heat bitch, to a fallen kid, to a rusty swing post was fair game. But it was a credit to Zippy's good nature that he disdained the role of mad-dog rapist in favor of the wooing, melancholy approach of the great romantics. He whined away warm summer nights whenever a dog was in heat within ten miles, his eyes soft with pleading. Unlike his feline counterparts, he seldom wore the battle scars of the roaming warrior lovers. He must have been pretty good at getting out of the back window in a hurry.

While I doubt he intended it so, Zippy gave me a vivid object lesson in the perpetually confused arena of sexual behavior. His fantastic prowess in that regard, which he seemed to have no trouble demonstrating many times a day, planted in my young mind the suspicion that maybe this was a quality that we tend to overestimate in a man. After all, Zippy was a dog. He urinated on the hydrangea bush, ate flies, sniffed the rear end of any other dog he met, and stuck his head into a bowl to eat. Even as a little boy I had firmly established within me the Creation order of value that decreed that dogs were different from women and men. Not much later in life, when I began to realize that a good chunk of our society and an even bigger chunk of our economy rested on the values of the "Zippy School of Sexuality," I had a constant reminder of the absurdity of those assumptions: I had only to picture the face of that old dog. I once heard a story of a Civil War era congressman whose short stature and high voice made him the butt of jokes about his probable lack of sexual potency. Never at a loss for words, the fiery lawmaker would reply, "Sir, you pride yourself on a quality in which…a jackass [is] infinitely your superior." Or Zippy. Despite all of our advances in learning and sensitivity, and no matter how clearly we see and even laugh at the other Neanderthal practices from our ancient past, the enormously resilient myth that ties male sexual worth to sheer quantity is still very much among us. It is sobering and humbling to think that after five thousand years of recorded history, and perhaps five million earlier years of patient development by the Creator's hand, a good percentage of America's men (in particular) continue to pride themselves on a quality in which they are demonstrably inferior to ninety-five percent of the creatures in the animal kingdom. If anyone thinks this criticism unfair he need only watch the Sunday football beer commercials, all of which disguise their intent about as effectively as the advertisers on Saturday morning cartoons. Real men like their sex to be quick, frequent, superficial; a physical release with whomever happens along. But when I see those good-looking, all-American boys sitting around the bar, ready to accommodate the bevy of beautiful women perpetually on hand just to please them, I suddenly see Zippy's head superimposed on each of them, and half expect them to begin peeing on the bar and sniffing each others' rear ends. Actually, of course, the comparison is unfair. When it comes to their idea of sex, Zippy was infinitely their superior.

Zippy seemed to grow old and die all of a sudden. Every bush for three miles around shuddered with relief. I was in high school at the time, and surprised myself with the sadness I felt at his passing. After all, he really wasn't our dog, and should have been far down the list of a seventeen-year-old's priorities. Maybe his death hinted at the separation I would one day have to face with Bubbles.

It used to be in the script of American boyhood that there had to be one great dog in his life. For Lloyd and me, Bubbles was that dog. She was the one who gave us our memories, showed us birth, and taught us that anything greatly loved required—though would not necessarily demand—great care. She stayed young and strong during our childhood years and then, as if sensing our lessening need of her, unobtrusively grew old when we reached our manhood. Bubbles was a boxer, nicely marked with a patch of white at the top of her chest amidst an otherwise unbroken sea of golden brown—except, of course, for the blackened face. She possessed the boxer's characteristically cropped tail and pug face, but we never had her ears clipped in the then-fashion of most bulls. Those floppy ears gave Bubbles much of her uniqueness. Lloyd and I would fold them up on top of her head (we could almost knot them) for the Rocky J. Squirrel look, or turn them inside-out to simulate the silly, tight hats worn by women in the fifties. But usually the ears just dangled alongside her head like open curtains around her face, easing the bluntness of her nose and rounding her eyes into soft pools of expressiveness. Bubbles was rather on the small side, though possessing typical bulldog power through the chest and neck. She tended to grunt air through her shortened nostrils when lifting her head to be petted, and left her short hairs throughout those parts of the house in which she was permitted. Mom, in her love for us, waited until we and Bubbles were gone before fully developing her allergy to animal hair. Dad brought Bubbles home from work one winter night in the middle-1950s tucked deeply inside his coat. We must have seen something of the game in his eyes, for I took the unusual step of diverting my attention from *The Three Stooges* (or whatever bit of inanity it was that robbed him of a decent welcome each night) as he stepped through the doorway. Without comment he set her down on the floor where she promptly displayed her literacy by squirting a perfect "S" on the carpet. No doubt Mom was the first to grasp the long-range implications from that initial performance, but if the thought bothered her she kept it well hidden, and pretended to be as happy about the prospects for this new family member as we were.

The second issue to be addressed, after sanitation, was a name. Because I was seven years old and enamored with the name of my new baby cousin, Marty, I suggested that we try something similar. "Farty" was the first possibility that came to mind, but Mom's giggling fit led me to suspect that the name was something less than acceptable. We settled on Bubbles after watching her bubble in her food during her first few feedings. But while Bubbles quickly outgrew her habit of bubbling in her food—her perpetually voracious appetite and no-nonsense attack on her food dish rendered bubbles a physical impossibility—she always remained well qualified for the name I had first suggested.

Bubbles, our good and gentle dog. *Photo from author's collection.*

Although she was a house dog, three of my favorite mental pictures of Bubbles were taken outside. One is of her romping through the woods and ravines with us on cool autumn days, racing up the steep hills then plunging headlong downward, sliding and tumbling through the leafy carpet while we were picking our way slowly along the rocky creek bottom. For every mile we put in during such treks she put in five. The only time I ever saw her stall was when she tried her full speed assent on "The Slate Mountain," a tiny hill so named by us because of its steeply ascending stacks of slippery shales. One of the laws of motion took her almost halfway up before she slid back, bringing an avalanche of slate with her. Unlike us, she remained unperturbed by that failure, just as she seemed unaffected by the stinging flies and mosquitoes.

Bubbles had a way of burning a tremendous amount of energy during her first minute or two after being let out of the house. As if hearing a starter's gun she would explode out of the front door and into a full sprint at stunning speeds. Dad, shaking his head, guessed that she was hitting forty-miles-per-hour in some stretches. That seemed a bit ludicrous, and I remember Roger Rutkowski's derisive taunts about our forty-miles-per-hour dog when Lloyd and I took Dad's estimate as gospel, but having once watched the greyhounds run I'm not so sure Dad wasn't at least half right. At any rate, we would have had no trouble convincing Billy Grocott of the fact. Billy, who would grow up to become a career police officer, was one of my best friends during my first few school years. He made the mistake of moving as Bubbles was making her flying turns after being let out one day. She never meant to hit him, and wouldn't have, had he stayed put. She just never figured out how to make late swerves at those speeds. Comparative sizes being what they were, she caught him somewhere between the knees and thighs. It took him quite a while to land. She did the same thing to another of my early friends, Gilbert Poulsen, who would grow up to own the last farm in Northfield. When we picked Gil up he had tears in his eyes, but he wasn't bawling. I guess the shock value counted for something. I don't remember that he ever came back. At one time or another, Bubbles got each of us kids while she was taking the checkered flag. Our yard was something like those "Little Rascals" finales, with people popping up into the air as they were clipped by the kids' various vehicles. She even nailed Dad one time. Then she met her match. Someone had, the previous evening, dug a deep ditch in the yard, perhaps for a utility line, but Dad had failed to associate this significant fact with Bubbles' high speed habit. After he let her out he winced as her streaking brown line intersected with that of the ditch. A dull thud and a minute later Dad saw her emerge slowly from the pit, blinking and shaking her head, wondering who had rearranged the earthen furniture.

The only other time we routinely let her out was, of course, for nature's calls. She had been well trained to take this task back to the woods line behind our back lawn. Long before we ever heard of Pavlov we had learned that Bubbles could do almost anything if she knew that an edible reward was in the offing, in her case a biscuit or dinner. She knew the routine so well, and was so anxious to get to the good part, that she would jam her face into the crack that appeared as we were opening the back door to let her out. The rest of her body contorted to get past the water softener that sat next to the door and blocked the very action she was trying to effect with her snoot. At that point, we would have to rein her back by the collar, usually lifting her front paws off the ground, before we could get the door completely open. Many years after she was gone, while visiting Mom and Dad, I absently looked down as I was going out that back door and saw a dark, boxer-faced smudge permanently smeared on both the door edge and trim, a monument to ten thousand excited exits.

Once in a while, her training failed her, especially at noon when she knew that her reward was not a dog biscuit but rather a full-blown meal. She also knew that the contract was, in this instance, ironclad: no output, no input. Sometimes, though, either because she was too impatient for her food, or because (we never considered this possibility) she did not have to go, Bubbles would start back to the house without having kept her end of the bargain. She probably could have pulled off the stunt if her conscience had not been so scrupulously honed. There were a couple of places out of our sight from the back door—the fireplace, a brush pile, and a few trees.

But after emerging from these well chosen spots her conscience immediately would betray her. If she sprinted back on the dead run, we knew she had done her duty, but if she came back in kind of an awkward, hopeful prance we turned her around and ordered another try. At night, the game required a sharper inference capacity on our part since much of both her business and her return trip were in the dark. One night, when Lloyd was having a good deal of trouble with her, he walked out onto the back patio and saw two little lights shining at him from around the oil tank where she had decided to hide for an appropriate length of time before reappearing.

This was a good and gentle dog. Dad said he would not keep a dog that snapped at children, even if provoked. Lloyd and I gave her plenty of opportunity. While never cruel, we could be masters of irritation (an art we practiced on each other more than her). Long before the Chinese told us anything about acupuncture we had located the scratch response spot on her side, and found considerable amusement in the bicycle-like response of her back leg. Probably the most common tease was a gentle tapping on the sides of her floppy jowls that we would do while repeating some silly line from a TV show or movie, not stopping until her playful growl would turn to agitated barks at which time Dad would tell us to "knock it off!" Sometimes, our games were more reflective of our boundless stores of juvenile energy. After watching Saturday morning "wrassling" Bubbles would become for us the reluctant opponent for the match that continued long after Lord Layton, the Gallagher Brothers, Ileo DePaulo, and the Flying Frenchman left the airwaves. After the ritual tapping of her snoot, just to get her sounding like a bad guy (Dad worked Saturday mornings), we would go through all of the badly contrived theatrics with her. The finale was the Atomic Drop, which one of us would effect by thrusting his neck under her belly, leaping to his feet (so that she now looked like a large canine neck brace), spinning around several times—all the while narrating the exciting action—then rolling her gently to the floor. The Atomic Drop must have violated every physical system built into her dog being—you could see the instinctive tension in her eyes—yet she never responded with more than a nibble at the hands that held her paws together throughout the ordeal.

The games that allowed her to stay on terra firma were less frightening but no less wearisome for her. Because she lived with two budding football players, Bubbles had to endure roll blocks to her front legs and tackles of her hind quarters (neither of which could knock her off all of her sure feet). But the worst stunt of all came the day we decided to set up a three-way test among her gifted physical attributes, her patience, and her love of food. After dressing her in one of our sweatshirts, such that the long sleeves extended six or eight inches over her front paws, we began sliding dog biscuits across the slick linoleum floor in the kitchen. The winner, hands down, was the food-lover in her, with her patience coming in a distant last. She insisted on sprinting after the skittering biscuits, a decision that turned her pursuit into a series of hard forward rolls into the lower cupboards. (I haven't the faintest idea how those cupboards survived our growing up years.) She seemed not at all upset with the price she was paying for the biscuits, but would flop her way back to us to see if her unbelievable good luck with these two nitwits could continue through a few more treats. Here were rewards that did not demand that painful trip back to the woods. What a steal!

But the secret to Bubbles' long-suffering nature with us was not training or discipline. She just loved people, pure and simple. Throughout all of our teasing her eyes never quite lost the look that reflected her pleasure at being included in that moment of our lives. She was a dog that never got sullen or jumpy, never shied away from us when we walked into a room, never spent time cowering in a distant corner. By all logic she should have been content to lie comfortably by the heat register, but she was always willing to trade that for a few moments of attention.

Visitors were treated with a gush of enthusiasm that bordered on the absurd. Because she had no tail to speak of, Bubbles compensated by wiggling her entire rear end, an art so poorly mastered that she often succeeded in folding herself completely in half. To this odd spectacle she would add a silly grin, repeatedly pulling her baggy, black cheeks up over her teeth. In her excitement she would also be snorting. I'm sure more than one first-time guest wondered if this folded, snorting boxer baring its teeth was rabid. But her greeting ritual went on whether or not anyone took notice of her. She would still be twisting, grinning, and sneezing long after our own greetings had given way to routine conversation. We would have to scold her away from the guest before one of her sneezes found its mark. She would then reluctantly lay down in her spot in front of the heat register, but kept her eyes riveted on the newcomer, and should his vacant glance happen to drift across her, the grinning would start all over again.

I think Neil White might have preferred rabies to Bubbles' curiosity. Neil was our minister at the Shepherd Road Church of Christ, a good man with a fine sense of compassion and not at all stuffy about his clerical dignity. Good thing. The last and most socially dangerous aspect of Bubbles' greeting of guests came if and when they seated themselves before we had a chance to get her settled down. This action would bring more interesting parts of their bodies into range for her communicative nose, an invitation she gratefully accepted after having to satisfy herself with feet, knees, and an occasional hand for the first several minutes of the visit. Visitors who were dog owners were not at risk—they instinctively knew how to intercept her with their hands or else quickly cross their legs—but Neil did not own a dog. He sat down on the comfortable chair and allowed his knees to flop wide open. Bubbles took the cue. I remember thinking, as the unthinkable scenario began to unfold, that this thing could not be happening to us. Not with our minister. Surely she would sense that this was a special situation, that here was no neighborhood kid waiting for a softball game. But if Bubbles flunked her test for discretion, I at least had to give her high marks for composure. Her poise was almost matter-of-fact as she moved in to get to know the real Neil White.

Not so innocently dismissed was her proclivity for vandalizing the house when the family left her alone for more than an hour or two. I always thought this bad habit was forgivable since I considered it her only fault. But it was a rather disastrous exception to good behavior, one for which many pets have been shipped back to the minors. Besides, it was easy for me to be forgiving since I was not the one held accountable for the household. I wonder now how many of Mom's evenings out saw the last hour ruined by her worry over what she would find when we returned home. Even I can recall some of the thick tension of those bleak homecomings. Still, none of us was willing to jump to the conclusion that she was a bad dog. Mom and Dad, normally no-nonsense types when it came to issues of behavior, hedged a bit in this instance, explaining to us that Bubbles' house trashings were caused by feelings of neglect, anger or fear. Of course, that never stopped her from getting a good thumping from Dad.

We first suspected that our dog had "a problem" along these lines when, while still a pup, she chewed her way out of a wicker hamper while we were gone. It was hard for us to believe that such a little puppy could tear through such tough stuff. It was only a hint of the developing power in her bulldog teeth and jaws, and she was on her way to a criminal career. We tried isolating her in the back room, but she quickly learned that a strong nose was as good as a hand in opening a sliding door. Then Dad played his big trump card, fastening her collar to a short chain nailed into the floor of the utility room. Lloyd and I felt sorry for her, standing there looking like some kind of advertisement for Devil's Island. We needn't have bothered. Neither her collar nor the chain could withstand her determined thrusts. She snapped both of them, at one time or another. Dad's last recourse was to wedge the sliding door shut and hope for minimal damage to the back room.

Actually, the last option worked rather well. We should have guessed it sooner, but there was a method to Bubbles' madness, and it had to do with her passion for food. I don't recall that she ever chewed up furniture or other fixtures, unless they happened to be in the way of where she wanted to go or what she wanted to get. The real action began after she clawed open the door to a lower cupboard. Typically, the feast might include a culinary array that would gag a goat: a half pound of brown sugar, part of a cereal box, a bag of marshmallows, some ground coffee, and a few tongue slaps at the flour. The resulting mess was unbelievable, but probably nothing compared to the inside of her stomach. Later in life, she finally went on the wagon and stopped her evil ways, for which she was allowed free rein of her parts of the house (the bedrooms and living room were off limits) during our absences. Then, very late in life and after many years of good behavior, she did it one more time. It wasn't a big mess—certainly not up to her former standards—just a few chewed open boxes and packages and some interested nibbling here and there. This time there was no beating, just headshakes and a few nostalgic smiles. Lloyd took a picture of it. In it she looks tired and a bit guilty, though I don't know whether from the deed, itself, or her weak performance of it.

Bubbles began to succumb to the calling of mortality just about the time Lloyd and I were succumbing to the feeling of immortality. We were in our middle and late teens, those years when death is, if not an impossibility, at least an absurdity. We could see the signs easily enough. She could not keep up with us so well anymore, and her black snoot was softening with white whiskers. Lumps were springing up on her body, a dull mist seemed to cover her once-bright eyes, and instead of sitting squarely on her haunches she would sag awkwardly on one hip. Sometimes her jump up to the back patio when coming in for lunch would impale her on the concrete edge, until she at last accepted the inevitable and began padding up the steps like the rest of us. But there was not yet any such inevitability in our thinking. She was still Bubbles, still chronologically younger than we, still the possessor of a voracious appetite, still an extension of the puppy that had peed an "S" on the living room floor and chewed her way out of the wicker hamper. So, we insisted on the pretense of earlier days.

One warm spring day of my junior year in high school, I decided to favor her with a walk in the woods. I no longer got back there as frequently as I had as a child, but something—perhaps my loneliness for Lloyd who was in faraway Tennessee for his freshman year in college—beckoned to me in the old manner. I did not even hesitate in taking Bubbles along with me. Yet somehow, the woods was not the same. Nature's badly strewn furniture was still in evidence in the ravines and creeks and moss-covered rocks, but the new leaves shuddered uneasily as we passed, as if peering over my head in search of a more appropriate juvenile guest to scramble up the slate hills, upend rocks in search of salamanders, and swing on the ratty-looking "grapevines." The visitor was courteously greeted and adequately tolerated, but no longer embraced.

I noticed that Bubbles was reacting differently, too. Instead of racing up and down the ravine walls she tended to stay on top of each ridge for as long as it ran, picking her way slowly to the creek bottom only if the topography left her no other option. She was walking rather than running, and even at my unhurried pace she was forever falling behind. Too, she was more prone to drift off in slow pursuit of some interesting scent or other minuscule curiosity that in former days she would have raced right by. Re-viewing that scene today I could better appreciate that one of the gifts of aging is the increasing capacity to see all manner of fascinating and beautiful things unseen in our feverishly hurried youth. On the one or two rare occasions when Grandpa Knowles accompanied us kids into the woods, we were frustrated by his methodical picking over a piece of ground we usually covered in two seconds, calling us back to see some obscure penny offered up from nature's vast largess. He was endlessly infatuated with wildflowers and lichens and animal tracks, while we wanted to know where rivers led and whether the semblance of a path was the beginning of the illusive Stagecoach Road or an old Indian trail.

Increasingly that day, Bubbles drifted out of sight. That irritated me. Sometimes she would belatedly respond to my loud calls and whistles, emerging atop the ridge without bothering to look at me, but other times she would not come, and I would have to go get her. Since I possessed a very shrill whistle that even a half-deaf dog could not miss, I firmly believed that she was using the beginnings of her deafness as an excuse to ignore my calls. The game of cat and mouse was not to go on indefinitely. Driven by my interest in moving down the ravine and my aversion to interruptions, I let her get too far away from me. I lost her. After twenty minutes of futile searching and calling, the implications of my careless indifference began to twist an icy knot in my stomach. Anger dissolved into fear. Then, for the first time, I began to realize that all of the possibilities seemed to work against her. She was an old dog lost in a vast tract of woods in which a dozen streams, a hundred hills, ten thousand trees, and a million rocks all looked the same. If she yielded to the more comforting lure of gravity she would proceed down the ravines toward the Cuyahoga, and away from home. Through it all, she would have to rely on cataract-dimmed eyes and noise-dulled ears, and even that assumed that nothing worse had happened to her in the meantime. What if she had fallen down one of the steep hillsides? What if Ruffie's ancient chows or their descendants had caught up with her?

That marked the first time that I consciously remember offering up a "bargain" prayer to God, making him a behavioral promise (that I ultimately broke) in exchange for her return. Then, I did the only other thing I could think to do; I went home. The weight of the world lifted from my shoulders when I saw her patiently sitting in the yard. An ancient instinct caused her to cringe slightly, but the only response I had for her was a grateful hug. I thanked God profusely, then forgot about him in my wonder at how she had managed to find her way home. I considered the homing instinct grandiosely trumpeted in the *Lassie Come Home*-type stories, but somehow that did not seem to fit. I was probably closer to the truth in my prayer, though my sequencing may have been a bit off. Many millennia ago God's creative touch was anticipating Bubbles' need when it created a sense of smell we weak-nosed humans cannot possibly understand. It was her one sense that did not wane in later years. In this case it had been sufficient to save her from the ignorance of a teenage boy, but it would not always be enough.

Sometime later, perhaps that next summer, Lloyd and I took her for a long walk along the line of the railroad tracks and into the large fields behind Virginia Smith's home, where in younger days we had picked some of the biggest wild blackberries in northern Ohio. For the two of us, it was a pleasurable trip into our past, even in the heat of a summer's day. For Bubbles it was an agonizing test of endurance, one that she would not pass. In her own quiet way she was trying to tell us that she was in trouble. She lagged behind, panted excessively, and in virtually every other way reflected the burden that was upon her. Not once, however, did she whine or cry out; it was not her way. We believed that she was just being stubborn and overly fastidious, and chided her accordingly. Once, she disappeared altogether, causing an irritating interruption and backtracking. I finally found her lying down nestled among some bushes on the shady side of the railroad embankment. Anyone not completely engrossed in his own world would have seen in a moment that she was struggling, and that those few moments of rest were a desperate attempt to conserve the few remaining physical resources she instinctively felt draining out of her aged body.

To my eternal shame, I shouted at her in my anger. For one long moment she just looked at me. By right of every natural law in God's well-considered universe she should have disobeyed my command and stayed put. But her eyes did not reflect the grievous injustice. Years of patience were in those eyes, patience forged in the fires of little boys growing to manhood in the sometimes rough, sometimes ignorant, sometimes selfish manner of boys on that journey. There had been hundreds of other injustices to her along the way, but of course she never saw them that way. Dogs seldom do. Slowly, she pulled herself up, awkwardly propping herself up on her front legs before pushing her bent hindquarters into a standing position. She made it back across the sun soaked field to a line of trees before her flesh finally failed her spirit. We were walking single file, of a sort, with Lloyd in front, Bubbles next, me last. It was probably poetic justice that I was the one to see her go down, to have the brutality of her crashing fall seared into my memory. Lurching suddenly to one side she heaved clumsily to the ground on her side. As soon as she hit the ground some secret shot of misplaced adrenaline caused her paws to scratch out a quick series of would-be steps, turning her around in a little semi-circle in the dry weeds. Then her body assumed an immediate rigidity, as if in the midst of the kind of deep, leg-stiffening stretch that dogs will do. A furrow creased her brow that only seconds before had been drooping heavily like the rest of her baggy facial skin. Most alarming of all, she had ceased her heavy panting. Her mouth was, in fact, clipped tightly shut.

 I yelled for Lloyd in a voice that must have seemed strange to him. It brought him to her side in a second. For a moment we stood in a stupor of disbelief and helplessness. Her eyes were sharply alert with the sense of crisis, but every other part of her body was being held in frozen hostage to the moment. I knew that she was in deep trouble as long as her mouth was shut, that her body, which had been gulping great quantities of air only a moment before, could not long exist on the meager supply being pulled through her tiny nostrils—assuming she was breathing at all. The moment may have lasted five seconds or fifty, but probably somewhere on the lower end of that range. Then, with an enormous sense of relief, I saw her powerful jaw muscles relax slightly and the tip of her tongue slip slowly through her black lips. In another few seconds she was panting again. The familiar, patient softness returned to her eyes. The summer-stilled tree leaves whispered lightly as the breath of death passed above us and through them.

In our relief we hungrily searched for a way to expend the futile energy that had built in our bodies as we watched Bubbles struggle for survival. We vowed she would not take another step until we reached home. It was a backbreaking vow. Her body was heavy, perhaps forty to forty-five pounds, and very compact. There was no way to share the load while we walked, so each of us had to cradle her to his chest, ala the husband carrying his new bride over the threshold. It was not possible to go more than one hundred to 150 yards before one of us had to hand her off to the other. It was the last hike she would take with us, and she seemed a bit embarrassed, or at least bemused, to be coming home in this way. Stepping inside the front door Lloyd set her down on the floor, unconsciously reenacting Dad's first introduction of her to us so many snowy winters before.

We were spared the pain of having pushed our dog into the throes of her death. She lived for three or four years after her stroke, and seemed to have suffered no permanent damage from it. As so often happens with America's outgrown pets, Bubbles' time of extreme infirmity and death fell on Mom and Dad rather than Lloyd and me. We were both at Milligan College by then. Lloyd, who was doing graduate work at the next-door

Emmanuel School of Religion, told me the news one day in the crowded halls between classes.

"Mom and Dad had Bubbles put to sleep," he said.

That was all. We looked at each other awkwardly for a moment, and then glanced away at the host of convenient distractions. Maybe he chose to break the news in this busy place knowing that I was not likely to do something silly. If so, he needn't have bothered. The time for anguished and desperate prayers was long past, and we both knew it. We had, at last, learned the final lesson that she would teach us, that of mortality. In fact, it was something of a surprise to us that she lived quite so long. Before each of our return trips to Milligan following breaks and vacations, Lloyd and I would each take a quiet moment with her, reasoning that it would probably be our last. Near the end she would not even get up to see us off, though we could read her heart by the slight wiggle in her stump of a tail. Separated by hundreds of miles and a new life, the prospect of Bubbles dying—unthinkable in the ravine and horrifying in the blackberry field—was now acceptable to me.

Not so, it seemed, for the people who had brought this gentle lesson-giver into our lives. As Mom said, Bubbles had been her daily companion for almost all of those fourteen years. While we were at school and Dad at work Mom and that dog were logging more daily hours together than any other combination in the family, canine or human. Nevertheless, I believe (to the extent I was able to read between the lines) that Dad was more affected by Bubbles' passing than any of the rest of us. This very pragmatic Depression kid, who had been forced to make sense of the loss of his mother at four, his home at fifteen, and his freedom and safety at eighteen, could not quite convince himself that a quick, painless death made sense for a dog mostly blind and deaf, crippled by arthritis, and bulging with cancerous growths. Years later he was still shaking his head and commenting on how she had never lost her lusty appetite. Maybe that was it: maybe he thought she should have been allowed to live until she voluntarily yielded that most boisterous of her life signs.

Or, maybe it was because Dad was the one who held her when the vet gave her the shot. She had come to us from his arms, and there she died.

Notes

1. Phillip R. Shriver, "The Beaver Wars and the Destruction of the Erie Nation" In Ohio's Western Reserve: *A Regional Reader*, edited by Harry Lupold and Gladys Haddad, 13. Kent: The Kent State University Press, 1988.

2. Lupold, *Forgotten People*, 32.

3. Lupold, *Joys of Pioneering*, 104.

4. Lupold, *Joys of Pioneering*, 105.

5. Lupold, *Joys of Pioneering*, 104.

6. Bessie Gooseman, A History of Olde Northfield Township (Northfield, Ohio: Historical Society of Olde Northfield, 1973), 11.

7. William Henry Perrin, J. Battle and W. Goodspeed, eds. *History of Medina County and Ohio* (Chicago: Baskin & Battey, Historical Publishers, 1881), 608.

8. Perrin, *Medina County*, 609.

9. Perrin, *Medina County*, 610.

10. Perrin, *Medina County*, 609-611.

7
An Odd Brew
The Reserve

Town square. Brecksville, Ohio. Nowhere is Connecticut's signature written more clearly on the face of the Western Reserve than in the many town squares dotting the area. The classic New England settings were as close to back home as many of the original pioneers would get. *Photo by Bob Vaughn.*

The unique history of America's most bizzare land transaction • The Connecticut Western Reserve • Settling "The Black Forest" • The ageless battle with the bottle

The settlement of this nation, while often lacking in benevolence, seldom wanted for economic logic. People gravitated to lands because they were located at the confluences of rivers, on the shores of natural harbors, or on plains well drained by river valleys. Since most of the land was to be occupied by people who would farm, they would scratch the surface to see if the glaciers had been generous in leaving some of the rich layers of loam created in the vanguards of their enormous, slow-moving prows. Sometimes a waterfall would catch a settler's calculating eye, giving him a vision of a grist mill or textile shop, or perhaps it would be a natural spring, gushing forth its obvious implications for people and livestock. From these natural advantages would grow small communities of people, struggling to survive by playing their one geographic trump card into a life of trade, milling or farming. Perhaps, if a particular village was blessed by good leaders and a bit of fortune—such as a canal or railroad right-of-way—its small cluster would grow into a great city. Conversely, the backlands of these population centers, composed of clay-lined gorges or situated farther away from natural transportation routes, would be settled sparsely and slowly. Most of them would never draw much human residential interest, as the hilly forests of southeastern Ohio today attest. Still, underneath all of these settlement patterns lay at least some kind of basic empirical logic of self-interest: when you see a piece of unoccupied land that will be of advantage to you, take it! Then, take advantage of it. This, for better or worse, is the way much of the U.S. portion of the continent was developed.

Not so the Connecticut Western Reserve in northeast Ohio.

The settlement of Ohio's twelve northeastern counties makes for a strange tale. Instead of grouping themselves along rivers, which the Indians had sense enough to do, the first white settlers simply hacked their ways into the middle of distant malarial forests, where they plopped down in single family units, hundreds of miles from home in Connecticut, and sometimes a dozen miles from the nearest white-faced English-speaking resident. When Isaac Bacon and his wife became Northfield's first permanent, white residents in 1807, settling in along what would become Valley View Road near the present Eaton Estates, he almost went beyond the capacity of his own hands. With no help available, Bacon was faced with the problem of how to hoist into place the logs for the crude cabin he desired to build. Eventually, he had to engineer a system of skid ramps and ox power to complete the necessary task. His wife was later to complain that for six months the only white woman's face she saw was that which dully returned her glance in a pool of still water.

Settlements in the Western Reserve rose up not along the natural transportation routes—twenty years after Cleveland's founding at the mouth of the Cuyahoga on Lake Erie the place was still a disease-ridden village, the prospects for which were rated below those of the thriving community of Brandywine a mile to the southeast of our property—but rather in the centers of neatly measured, five-mile square plots. The settlers not only had not seen the land, most of them had never been west of the Connecticut border. Instead of dark looming forests and thick carpets of mosquitoes, their eyes had seen only inked squares on maps in Connecticut land offices. It was the worst conceivable method for conquering a rugged land. That the whole business somehow took root, survived, and then thrived, is one of the most interesting stories from America's frontier.

The basic problem lay with the preposterous land claims of some of the original colonies. The King's early proprietors had a pretty good idea of their north-south boundaries, stacked up as they were along the Eastern Seaboard, and the Atlantic gave them a mutually agreeable understanding of east. West was another matter. No one was sure how far the land extended in that direction. To be on the safe side the founding colonists got into the habit of claiming everything westward to the great south sea—whatever and wherever that might be. Once time improved their sense of geography a bit some of the colonies were agreeably disposed to the idea that the Pacific matched their original claim descriptions, even if it did mean that their manifestly destined states might take on the shapes of three-thousand mile long earthworms stretched across the North American land mass. This naturally caused conflicts as claims inevitably overlapped, and the smaller land-blocked Colonies complained that their western neighbors were getting too many desired pieces of the continental pie.

Reasonable people soon saw that the issue could never be settled in the nation's courtrooms, and a sometimes bloody border dispute that boiled up in the six years before the Revolution (and lingered until 1799) in which Connecticut men were killed by Pennsylvanians shocked many others into the realization that violence was also folly. The only rational solution was for the colonies to give up all western land claims to the new government under the weak Articles of Confederation. It was a grudging concession, but a rational one made by most of the colonies with remaining claims after Virginia took the lead. But Connecticut was not being rational. Perhaps sensing that the young nation could be softly blackmailed by the precariousness of its position (many of the European governments were giving the fledgling republic little chance for survival) Connecticut pushed for and won a compromise concession in the Ohio country. The farthest reach of this Connecticut Western Reserve, which would become Huron and Erie Counties, was designated as the Firelands to compensate Connecticut citizens burned out of their homes by British fire raiders during the Revolution. The remainder of the 120-mile stretch, filling the space between the forty-first and forty-second parallels back to the Pennsylvania line, was to be sold to private interests in order to create an endowment for Connecticut schools.[1]

Under the circumstances, which included Indian occupation (at the moment much of the Reserve land was inconveniently located on the wrong side of the nation's recognized western border on the Cuyahoga River), inaccurate maps, and forests so monstrous that even the Indians kept to the trails and rivers, it is not surprising that this great real estate venture often realized its great potential for disaster. The first task was the surveying of the land, beginning in 1796. During the ensuing couple of years, two separate parties hacked their ways through the brutal wilderness to complete what has to rate as one of the greatest surveying feats in history. In sweltering heat and under thick clouds of flies and mosquitoes, the men ran their chains through the jungles, relying chiefly on often capricious compasses and astronomical observations. They were tormented by malarial fevers, labor strife, communications obstacles, and supply problems. Equally troublesome was a board of directors back in Connecticut consumed by greed. Somehow the surveyors managed the overwhelming task, marking off the three million acres into twenty-four ranges, each containing from six to thirteen townships. Several of the surveyors imprinted the new land with more than the five-mile square lines that defined each of the townships. Their surnames read like a map of northeastern Ohio: Moses Warren, Joshua Stow, Joseph Tinker, and Moses Cleaveland; but several of those whose names never graced a township, names such as Spafford and Doan, were also instrumental in giving flesh to the skeleton they had so painstakingly measured.

Given the working conditions, their calculations were remarkably accurate. The troublesome north-south survey lines, sometimes bowing inward or outward slightly as they groped for Lake Erie, could give the townships' northern borders a smidgen more or less than their allotted five miles. Still, greater accuracy was not within the capacity of eighteenth century human beings. The total acreage was accounted for to within an error factor of less than three percent. Yet that was not good enough for the board of directors. Discarding maps that did not confirm their avaricious hopes, the directors doggedly pursued the possibility that the shoreline of Lake Erie ran more directly east-west than the northeast-southwest slope suggested by the map of Moravian missionary John Heckwelder. So great was the wish that fathered the thought that they irritably dismissed the heroic efforts of their own surveyors with hardly a thank you, then created the Excess Company to develop the hundreds of thousands of additional mythical acres envisioned in the windfall. When, ultimately, the land proved to lie on the bottom of Lake Erie, many of them lost their shirts.

The whole, sorry episode was a testimony to the motive that drove too many of the Western Reserve's speculative planners.²

One of the repercussions of the directors' poorly disguised greed was the violation of the border separating the United States from Indian lands. Many of the great Indian trails of the period had, for centuries, wound through sections of the Reserve. Most of these were east-west routes, shadowing the line of the lake, and not favorably disposed toward arbitrarily assigned and invisible north-south surveyor lines. In the days when the Indians were still winning a good percentage of the battles, Anthony Wayne won a surprising 1794 victory at Fallen Timbers in northwestern Ohio when the heavily favored Indians proved ineffective against well-used bayonets. The resulting Treaty of Greenville the next year had created a U.S. western border so irregular that anyone looking at the twisting line could see in a moment that it was not meant to be permanent. The top part of the line ran from the mouth of the Cuyahoga to the Portage Path in what would become Akron. Of course, this line along the Cuyahoga River sawed right through the Reserve with which the young government had satisfied Connecticut's western claims nearly a decade earlier, but those kinds of technicalities were of minor significance in late eighteenth century America. The circumspect Moses Cleaveland had carefully established his survey headquarters on the eastern side of the Cuyahoga's mouth in order to convince the Indians of his good intentions, but Excess Company investors soon forced his second-in-command, Augustine Porter, to drive west into the Firelands in a vain attempt to prove the existence of those nonexistent excess lands. There would be blood to pay for this greedy stupidity. Some of the Indians who observed this outrage had fought under Pontiac's banner a generation earlier, and others would again take up the hatchet one last time in the Reserve fifteen years later in wars under the great Shawnee chief, Tecumseh.

Under these tremendous handicaps, the Reserve's first settlers proved a remarkable breed of pioneers, probably more so than the more celebrated folk who first peopled the great western plains and mountains. Where many of the latter had always lived on the ragged edge of the frontier, and were used to privations, the Connecticut sojourners came to the heart of the Black Forest from comfortable homes in an area that had been settled for well over a century. Because wheat does not grow in the woods, the badly isolated settlers had to invest two or three years clearing the land before their dreams became even theoretically possible. Their journeys to these lands of deliberately exaggerated description were themselves high-risk ventures. The water route put the travelers in unseaworthy boats at the mercy of Lake Erie's volatile and capricious whim, while the overland trek through the Alleghenies traded those risks for bandits, con men, and Indians.

David Hudson got the worst of both worlds, rebuilding a damaged boat to carry him almost literally through our backyard via a Lake Erie/Cuyahoga River/Brandywine Creek route while sending his livestock and supplies overland with Benjamin Tappan, who was on his way to the founding of nearby Ravenna. After Hudson's first journey, during much of which he found himself "most heartily repenting ever having undertaken the expedition,"[3] Hudson returned to Connecticut, where he spent sleepless nights pondering the wisdom of taking his wife and family into an environment in which

> ...so many chances appeared against me; and should we survive the dangers in crossing the boisterous lakes, and the dis-

tressing sickness usually attendant on new settlements it was highly probable that we must fall before the tomahawk and scalping knife. As I knew at that time no considerable settlement had been made but what was established in *blood*, and I was about to place all those who lay around me on the extreme frontier, and as they would look to me for safety and protection, I almost sunk under the immense weight of responsibility resting on me.[4] (emphasis in original)

The Reserve's dangerously fickle weather gave the Hudson family another preview of things to come during the second trip in 1800. Whereas the journey of 1799 had been plagued by low water levels, so much so that Hudson had been forced off the Cuyahoga at Brandywine Creek, the return trip saw flooding in the wake of torrential rains, compelling the county's first white residents to spend five harrowing days marooned in the Pinery Narrows under what is now the Route 82 bridge over the valley.[5] But the resolute Hudsons would not let go of the fingertip hold they had established in the middle of this hellish nowhere, and because of that fact, tens of thousands of people, including many Knowleses, would live their lives differently.

Others also weaved their names into the fabric of the Reserve's tapestry. Captain Joseph Tinker, a boatman on Cleaveland's first surveying party, posthumously lent his name to the Cuyahoga's longest tributary after he and several others drowned during a violent storm. (Had we kids known he was a hero we might not have invented the inevitable nickname, "Stinker's Creek.") John Young, Turhand Kirtland, Eliphet Austin, Gideon Granger, and Nathaniel Doan (Cleveland's Doan's Corners) joined the list of name-givers although, as William Ellis has noted, permanent place names did not always directly reflect the value of the contributor.

> For example, though five distinguished Kelleys came and settled here, and operated extensively in land, not one town is named Kelley. On the other hand, John Young's town bears his name though he came for only a brief stay. Almost every town has a Granger street, yet Gideon Granger's contribution was highly localized.[6]

~~~

In my home town of Northfield, development was often, but not always, a microcosm of what was happening throughout the Reserve. The most noticeable difference was the slower pace of settlement. The tardiness was actually a compliment to the township. Of the approximately 200 townships created in the Reserve, Northfield, along with Bedford, Warrensville and Perry, were identified early as the four equalizing townships set aside to balance the luck of the draw for the other properties. These four were judged to be the most valuable, and so were separately categorized from the lottery draw in 1797. Because there was more interest in these prized township properties, it took longer for the deeds to get out of the hands of speculators and into the hands of people who actually wanted to work the land. Northfield's high value was accounted to its rich soils, and of course, its direct access to the Cuyahoga. Another difference lay in the development of smaller towns within that of the larger township. Perhaps because of the slow progress made at Northfield's center, outlying settlements at Brandywine, Little York, and

Macedonia grew into thriving community clusters at early dates, undermining the kind of township centricity David Hudson had instantly created when he planted his settlement in the heart of his five square miles. Even today, while some of her neighboring townships line themselves up in neat squares, Northfield is cut up into Northfield Center, Northfield Village, Sagamore Hills, and Macedonia.

Like the larger Western Reserve, Northfield Township is filled with roads and creeks and corners named for people from the long ago. The Historical Society of Olde Northfield is home to an early township map reflecting a time when the names on it referred to real people, not just places. In the mid-to late-nineteenth century, when the vast majority of Northfield's landowners were farmers, men like Comford Chaffee, John Nesbit and Dwight Boyden each owned a piece of land that now provides room for several hundred residences on a myriad of short streets and cul-de-sacs. Our little loop, somewhat older than those others, could (and can) only be accessed via South Boyden Road. The three farmers are long gone, but three roads bear their names. Other names from Northfield's nineteenth century probably never saw their names on the township maps. Too bad, for some of them might have added considerably to the joys of map-reading. Resolved Palmer, Shepherd Honey, Consider Taylor, Roby Scriptures, and Pardon Brooks were people from a time when names were worth more than a sound.[7]

I am endlessly fascinated by those old names, those earlier lives. Part of my fascination is driven by the concern that they are so faded from our memory. As recently as the mid-twentieth century, it was not unheard of for someone to devote months or even years of their lives to resurrect the images of the saints and sinners who walked Northfield's distant past.

Several late nineteenth century Summit County researchers (Perrin, Lane, Bierce, Wickham) went to the trouble to dust off the lives of the township's early settlers, and Bessie Gooseman completed a short, book-length work in the late 1950s (*A History of Olde Northfield Township*). But such civic sentinels are hard to find now, and the leads are growing colder. I can vaguely remember Mrs. Gooseman as a substitute teacher who occasionally worked a day at the Northfield Elementary School when I was there. Because I was, maybe, ten, and she was somewhere between sixty and one hundred and twenty, I of course assumed that she probably had nothing better to do with her sunset years than knitting, petting the cat. If someone had told me that she was working on a history of the township, I would have nodded knowingly; that kind of thing seemed to fit her, too. Mrs. Gooseman's *History* is not really a history, but rather a pie-filling of names, dates, and unedited reminiscences from some of the old-timers. Yet, the passion for the past is there. That ten-year-old boy would have been shocked to learn that by the time he became little more than half her age he would be driven by the same fires that burned within her.

Other names on the old map awaken images from my school years in the 1950s and 1960s. Chuckie Ozmun, Johnny and Keith Holbrook, and Elizabeth (Betty Lou) Bliss each bore names that appear on that map, though I don't know if these were direct lineal connections. Chucky and Johnny, both a year and two older than me, were fun-loving types, while Betty Lou, who faded from my class picture sometime around elementary school age, was a round, shy girl who would not even make noise when she laughed. In third grade, Mrs. Oviatt hauled a friend and me out into the hall one day and favored us with one of her finger-waving lectures after we made fun of Betty Lou's seemingly early bust development. It was probably a bit premature of her since Betty Lou's figure was guilty of little more than baby fat, but Mrs. Oviatt went the whole nine yards, awing us with a story of a woman who had lost her breasts in a fire. I was so mesmerized by her account that the shock did not wear off for a couple of years, by which time girls' figures were, for me, no longer a laughing matter. I don't know if Betty Lou was directly connected to the old Blisses, but the name was one of the township's earliest and proudest. Ambrose and Lucien Bliss owned a big chunk of the center of Northfield in the mid-nineteenth century, the latter offering a piece of it for a part of the town square, and Ellen Bliss, as chairman and historian of the Northfield Committee, supplied the history of the township's women later in the century. When Charles Whittlesey, Cleveland's famous geologist and anthropologist, excavated the old Indian fort (Number 5) on the ridge just north of Red Lock Hill, his notes say that he did so with the help of Lorin Bliss of Northfield.[8]

Mrs. Oviatt herself wore a name rich in the Reserve heritage. It was a name that rode in with the first of the valley's pioneers, connecting itself with the Hale family in Bath as well as David Hudson. On the old Northfield map, two Oviatts appear close to the property where my widowed teacher lived into her nineties. Originally that part of town was known as Little York, a thriving community far more significant than Northfield Center. Little York was located at the confluence of Brandywine Creek and Indian Run, about at the spot where Highland Road (known to the old timers as Little York Road) intersects with Old Rt. 8 (Akron-Cleveland Road). There is nothing there now to indicate a once thriving settlement. Interstate 271 slashes across Old 8 at the spot, rudely burying the creeks and slicing Highland into unjoined halves as it hurries toward Cleveland's upper east suburbs. Mrs. Oviatt's old house looks like just one more fading residence on a small lot. She hinted that the interstate's planners probably helped her husband to an early grave, using the knife of eminent domain to cut off the bigger half of their land, including the well, then tossing a few dollars over their disappearing shoulders as they pursued the promise of progress. Even by then, however, Little York was already only a dim memory.

A mile or so down Brandywine Creek from Little York is another interesting settlement ghost. In 1816, Brandywine was considered by some a brighter urban prospect than Cleveland.[9] Located halfway between Little York and the Cuyahoga the settlement grew naturally around one of Northern Ohio's most magnificent waterfalls. There is dispute whether the name honors the Revolutionary War battle or a favored rum of the time, but no disagreement clouds the high historical value of the meandering stream. Yet, like Little York, Brandywine has become a communal nonentity. The same Interstate 271 that vivisects Little York's neglected grave also slants garishly across the backdrop of Brandywine's once majestic stage. But unlike its upstream neighbor, Little York, the Brandywine settlement has managed to cling to an identity of sorts. The Cuyahoga Valley National Park constructed lengthy, wooden walkways into the south wall of the canyon adjacent to and below the falls, and the remnants of the last factory driven by its water power are carefully preserved at the top. The house that stands vigil near the falls, which we as kids knew to be the home of friend Tuck Richards, was beautifully restored into one of Ohio's finest bed-and-breakfast lodgings in the early 1990s.

The first sign of permanent white life at Brandywine came in the person of George Wallace, who quickly converted the waterfall's power potential into a sawmill in 1814. His brother, Robert, followed with his family the following year, at which time a gristmill and distillery were added to the fast-growing center. A store and a woolen factory followed, all of which required the labor of twelve to twenty employees, a veritable industrial complex by the wilderness standards of the early Reserve. Perrin also notes that Northfield's first school term was probably taught at Brandywine in a log cabin constructed for that purpose in 1817. The bustling settlement had become so important that by the time the Ohio Canal came plowing through a mile to the west, the township would not hear of allowing it to remain in Boston, its legal home. To this day, Northfield's southern line is interrupted by a rectangular dip that reaches down to claim the Brandywine prize.

I was surprised to learn that the southwestern part of our township was among the earliest settled. During my growing-up years the houses along Highland Road looked more like rural retreats than planned subdivisions, and the hills sloping down to the Cuyahoga suggest an uninterrupted, primeval parade of wilderness. But that is a deception. Cattle must have grazed on those hills—as those bits of barbed wire buried deeply in the trees behind the ball field suggest—and the land was probably not much quieter than it is now. Gooseman hints vaguely that a variant of the Mahoning Indian Trail looped over to the foot of Red Lock hill before heading back north up the old Holtzhauer Road extension, a path not noted by Wilcox and the other Indian trailers, but one that also makes sense given the location of Pontiac's boyhood Ottawa Village in the valley. At any rate, it is known that Highland (Little York) Road, crossing behind our back property line, was a busy thoroughfare at a very early date as it carried trade from Little York's sawmill to the lumber-hungry canal at Red Lock.[10]

I still wonder about those voices from that elusively vibrant past. Who were these people who quietly scratched the same ground over which Lloyd, Jack, and I so casually romped a century later? If not our blood ancestors, they were none-the-less brothers and sisters of the land, a not-insignificant familial bond in a world of fortressed acquaintances. Surely, these geographical forerunners were once filled with the worries, hopes, joys, and agonies through which all of us wade during our journeys back to God. Each, in his or her own way, must have made for a fascinating story.

Unfortunately, fascination sometimes prefers to work on the dark side of the human soul—then and now. I found that my boyhood soil contained not only arrowheads and canal clay, but also the blood of Northfield's most notorious murder case.

On the early map, our little homestead was contained in Lot 69, owned by J. and J. McKisson. One small farm to the north was another large lot, number 79, owned by Marlin McKisson. Since the map dates to the latter half of the nineteenth century, I make the assumption that these folks are a couple of generations removed from the "murderous McKissons" of 1837. The McKisson name is infamously burned into Northfield's corporate memory. Physically the family accounted for a huge stretch of ground between Brandywine Creek on the south and Valley View Road to the north, with only the small plot of Lucian Bliss interrupting the flow just south of Rt. 82. Robert McKisson's north-south finger of land extended north of 82 for the better part of a mile, covering at least some of the ground that would become the Polsen farm. The names so matter-of-factly penciled on the map sometimes are thin coverings for traumatic lives. One warm, summer night in 1837, Robert McKisson's wife was brutally ax-murdered as she lay in her bed. This grandmother of a later mayor of Cleveland did not immediately die after the blow, creating a macabre scene in which neighbors actually spoke with the half-headed woman laying amidst the splatterings of brain and blood. As if the family did not have trouble enough, Catherine McKisson stated that the deed had been done by her father-in-law, Samuel McKisson, a testimony confirmed by her daughter, Lucinda, who had suffered the blunt end of the ax seconds after the fatal blow. Although David McKisson, another son of Samuel, seems not to have been mentioned by either of the women, much of Northfield's suspicion accrued to him, especially when someone remembered seeing him running from the direction of the Robert McKisson home toward the canal the night of the murder. David had been making a nuisance of himself for several days before the murder in trying—apparently not without some success—to win the affections of Lucinda. The community reasoned that Catherine McKisson's active resistance to the match had driven David into the murderous rage. That supposition was born out when he was arrested in downtown Toledo with dried blood on his clothing. He was tried, convicted, and, in February of the following year, executed, but, like many old murders in rural areas, this one left questions that wouldn't go away. Why was Old Man McKisson acquitted of the same charge when he had been named by both victims, one of whose statements, as Perrin gravely notes, "was the dying declaration of the murdered woman?" And why would David McKisson bludgeon the objects of both his hate and his love within the same house? If the young lovers had exchanged rings, as Perrin asserts, the resistance of a sleeping woman should not have seemed much of an obstacle to their plans. All of which proves that even in little Northfield the soap operas were running long before television.[11] A half-century later, Summit County writers were still buzzing about the sensational aspects of the sordid case, but Ellen Bliss, historian of Northfield's women's committee, would not degrade her history of the township's pioneer women by mentioning the details of the "sad circumstances attending her [Catherine's] death."[12]

Like the name McKisson, that of Viers was dotted all over the early Northfield map. Just south of Little York, a mile and a half from our homestead, was the lot of Caroline Viers, while the same distance to the north found the property of Charles Viers. There were others as well, but it was the name of Dorsey Viers who made Northfield a household word in the Western Reserve long before Cryus Eaton or the Northfield Racetrack. According to Lucius Bierce, never one to play down the sensational aspects of a story, Viers was as good as convicted of murder in the minds of Northfielders after Rupert Charlesworth mysteriously disappeared from the Viers property one night in 1826. Charlesworth, a native Englishman and something of a rogue (the constable had discovered his disappearance while going to arrest him), had been boarding with the Viers family at the time. In the absence of a corpse the mounting circumstantial evidence began to appear overwhelming. Viers had told conflicting stories to the constable, someone remembered hearing a rifle shot at the Viers cabin on the night in question, a hired girl on the Viers farm testified that she had found Charlesworth's blood-caked blanket concealed in a haystack, and—probably most damning in the cash-starved Reserve of 1826—the dirt-poor Viers family was suddenly awash in money. As a final touch, a human skeleton was unearthed in the nearby woods.

Between the "murder" date in 1826 and the trial date in 1831, work breaks in Northfield's farm fields saw animated discussions affirming the growing conviction that the township was harboring a murderer. When

Justice of the Peace George Wallace finally got the indictment before him, a sensational trial of eight days followed. The result was the most aggravatingly unsatisfying outcome imaginable. With the prey virtually in the teeth of justice, two strangers from Sandusky suddenly appeared to say that they had seen Charlesworth alive and well long after the alleged crime date. Coupled with the lack of absolute proof of the Englishman's demise, this new and unwelcome testimony killed the prosecution's case. But having invested five years in meaningful gossip, few people in Northfield were going to allow themselves to be fooled by this ruse. Obviously, Viers had hired the two to perjure themselves, then leave town with nothing lost and probably something gained. In those times, Sandusky was as good as a million miles from Northfield.

The point was not lost on Dorsey Viers, and therein lies the fascinating part of the story. Having already gone through five years of societal hell, and facing similar prospects for the remainder of his days in Northfield, Viers began searching the corners of the young republic for Charlesworth in the naive hope of exonerating himself. In Viers's day, citizens often had to hunt down their criminal offenders when the law would not. But Dorsey Viers went far beyond the requirements of even this crude standard. Bierce says that he spent years in lonely wanderings and strange taverns, always asking the same question, always getting the bemused stares.

> He visited all parts of the Union, and, after a search of years, he one day went into a tavern in Detroit, and in the presence of a large assemblage of men, inquired if anyone knew of a man named Charlesworth. All replied no. Just as he was about to leave, a man stepped up to him, and taking him to one side,

inquired if his name was Viers, from Northfield. Viers replied that it was. The stranger then said, "I am Rupert Charlesworth, but I pass here under an assumed name."[13]

Charlesworth must have been much moved by the plight of Viers to have so endangered his fake identity. His legal problem, and the one for which he was being pursued by the Northfield Constable on the night of his disappearance, was that he was probably too familiar with men like Boston Township's Jim Brown, the notorious king of the valley counterfeiters. Having none too subtly passed a bogus bill, he had skipped town for the friendlier skies of his native England until the episode blew over. Even after several years, Charlesworth did not feel comfortable using his real name in the United States, a fact that made his willingness to return with Viers to Northfield all the more remarkable. A town meeting was immediately called at which all but one of Dorsey Viers's suspicious neighbors conceded that the murder victim had remained remarkably well preserved during the ensuing years. It was one of the livelier moments in the Western Reserve's early history. A half a century later, Viers's family name was in good standing in Northfield.

Viers's pathetic wanderings would cast a prophetic shadow over the sensational case of Bay Village physician, Dr. Sam Sheppard, who was convicted in 1954 of murdering his wife, Marilyn. Sheppard—and later his son—would spend a lifetime trying to clear his name and find the real "bushy haired" murderer Sheppard claimed to have seen (events that inspired the creations of the wildly popular television series, *The Fugitive*, in the mid 1960s that saw the fictitious Dr. Richard Kimble on the lam, feverishly searching for the one-armed murderer who no one believed existed). Sheppard, after serving ten years in prison, was ultimately exonerated by the courts—and Kimble by the scriptwriters—but Dorsey Viers had to count on his own resources to satisfy his never-dismissed moral jurors in Northfield.

~~~

It might have been called the Whiskey Reserve. Across the river in Brecksville, Columbia Road was known as "Whiskey Row" because of the large number of taverns lining its way. Back over on our side of the valley the next lock up from Red Lock, below what is now Greenwood Village, was Whiskey Lock. The Wallaces were not unique in building a still on the heels of their gristmill at Brandywine. The pattern was the same in most places. George Condon states that Cleveland's first business was a distillery[14], and Lorenzo Carter's single log cabin on the west side of the Superior Street hill was licensed as a tavern at the same time it began serving as a school in 1802.[15] Even the religiously circumspect David Hudson included a stash of 31 3/4 gallons of New York whiskey on his journey to an already wild-enough western frontier. If Hudson carried that matter on his conscience he found relief at Conneaut when thieves made off with the supply.[16]

> The problem was that this most volatile of substances was also a recognized currency in the cash-poor west. Brandywine's distillery turned out thirty to forty gallons of "excellent whiskey" a day in 1887, an amount so far above the needs of the local population as to verify an extensive market. The old Wallace account books from Brandywine revealed that "…not only was

whiskey used by every one, including ministers (and perhaps abstainers), but was used extensively as an article of exchange, serving the purpose almost as well as bank notes."[17]

The Indians were well aware of the whites' remarkable capacity for producing hooch, a fact that played no small role in the story of the American West. While images of Indians' fondness for whiskey fall outside of today's politically correct sensitivities, there is simply too much historical evidence to be hidden under that rug. Allan Eckert's extensive Indian histories are filled with accounts wherein key decisions—often relating to life and death—were directly influenced by the presence or absence of the hard stuff. In Northfield, the wife of charter settler Isaac Bacon found herself confronted one day by six of the Indians who invited themselves in for a go at the jug they knew was kept in the cabin of every white settler. Sensing the obvious danger in that prospect, she told them she had no whiskey, but they would hear none of that. A quick search produced a full gallon of it, which they quickly killed in a rapid passing from one to another. When Mrs. Bacon tried to intervene, they drew their knives and made clear with motions what they could not communicate with words. After intimidating her for a time they finally left without doing any further damage.[18]

The Indians were not alone in their difficulties with whiskey. Too many whites were also under the influence of this frontier staple, and they, unlike their Indian neighbors who tended to do most of their drinking in fantastic but singular bursts, often made it a daily habit. I wonder how many vistas of Ohio's magnificent and unspoiled scenery in the early nineteenth century were bleary images viewed through bloodshot eyes. Thirty miles to our northeast at Kirtland, near the headwaters of the Cuyahoga River, Christopher Gore Crary took a closer look at some of the hardy pioneers of that historic town and turned up a fair number of lushes. Commenting on the quality of life in the town after the erection of a still in 1819, Crary concluded that the successful establishment of the still proved to be a pyrrhic victory for Kirtland.

> It made it convenient to get whiskey, but did not increase our home comforts. It made a market for corn, but did not increase our cash receipts. It brought in some inhabitants, but did not improve the morals of the place. It made some business for magistrates and constables, but did not promote peace…From being a blessing, as was hoped, the still-house became an unmitigated curse.[19]

Pouring over the distillery's old account books, Crary found entries for 138 persons during a fifteen-month period between 1831 and 1833. Twenty of these found uses for at least a pint of whiskey a day (assuming the Kirtland still was the only one they patronized), some as much as a quart a day. One man, whose job earned him fifty cents a day, would spend $5.22 of his $9.09 paycheck on drink before parceling out his remaining pennies for the necessities of his family. Clearly the frontier brew could do as much damage subtly as it could violently.

Crary's careful eye also noted the intensity of the Kirtland Temperance Society. Two hundred and thirty-nine members signed its constitution, and once they focused their corporate glare on the town's sore spot its days were numbered. Choosing the easiest plan of attack, the society bought the still house and purged it of its intended function. Somewhat ruefully, Crary noted that the society died under the heavy influx of the Mormons who established their headquarters in Kirkland in the 1830s. Apparently the Mormons took the term "temperance" at face value and not as a synonym for abstinence, as is the habit of many temperance groups. While offering grudging praise for their general temperance, Crary could not resist passing along a report that "they consumed a barrel of wine and other liquors at the dedication of the Temple, enabling some of them to see angels, have visions, prophesy, and dream dreams."[20]

America's life-long affair with alcohol has become an accepted relationship in the nation's familial fabric since Prohibition's dismal failure almost a century ago. The mistress has been subtly brought into the house and given a bedroom and place at the table. You don't hear much about temperance groups today, and the term "teetotaler" has run a pejorative course back into the dusty images of a time when men and women thought alcohol an enemy worth fighting. We even kid about our drinking now, in something of the tone we might use on a beloved aunt. In the course of my work life I would attend one or two national conferences each year on how to fight substance abuse. In the midst of the familiar litany of the evils of "crack," "ice," and other drugs, one of the early morning speakers would invariably make a joking reference to how he and someone else at the conference were not at their best that morning because they had "closed down the bar last night," or how so-and-so "looked a lot more subdued this morning than when she was dancing on table tops last night." The audience always laughed easily, as if a minister had just told a safe joke while warming up for his sermon. There, among some of the most knowledgeable crime researchers in the nation—people with piles of data that confirm on a daily basis the fact that alcohol is still America's worst drug of abuse—the foe finds curious comfort.

Maybe the thing is inevitable, irresistible. If I sound too fanatical about this "foe," I should confess that an occasional social drink (I'm already retreating to our large stash of forgiving language here) is not beyond my reach. "Drinking" is a curious term, forever frozen in the present tense, as if announcing to the world that it will never go away. Among the world's earliest historical records are evidences that people had already learned how to ferment a fruit or grain into an intoxicant. No doubt Eve enjoyed eating that first piece of fruit, but I suspect that the second one went into a still. It is as if there was an understanding from the beginning, a deal, that if we would agree to try this rugged walk called life, God would wink at our tendency to make some crutches along the way. Sometimes his hand seems too far away to reach, so, for those times…The pioneers of the Western Reserve must have often felt the need to invoke the terms of this kind of deal. The Black Forest was immensely lonely, the odds against making a life here were fearfully long, and there was that thing back east, that horrible happening or person or debt that they were trying to forget.

~~~

The Black Forest was only an ancient memory by the 1950s, but men and women still needed the chemical crutches. Physical isolation had ended—people were piled on top of each other in Cleveland and Akron—but the loneliness remained. An icy existentialism covered the walls of the bomb shelters and the privacy fences as people came to the horrible realization that neither their own lives nor that of the planet was controllable any more. The Utopian hopes that still burned brightly just two generations before were deeply banked under the presence of the dark, sinister evil that lurked behind that wall in Europe, that skeletal dome in Hiroshima, and those nameless crosses in France. The marvels of liberating technology, about which we were forever congratulating ourselves, somehow resulted in less freedom, not more.

Worse were the walls that went up inside all of those dream ranch homes that were suddenly cluttering the post-War American landscape. Too late, many of the returning GIs realized that the white picket fence, the good job, the wife and two kids and even the dog guaranteed neither the means nor the end in the pursuit of happiness. The nation had passed from childhood into her busy majority, and the world was proving to be an interesting place, but ultimately the whole business still came down, as always, to a solitary individual wrestling with his or her soul, haunted by the echoes of an increasingly avoided spiritual call, searching aimlessly for whatever it was that he, as a human being, was supposed to see, to be. In such a world alcohol was as welcome a companion as it had been 150 years—or 5,000 years—earlier.

I suppose it is fair to say that there was a drinking problem in our family. But I am not sure whose it was. Dad was the one who did the drinking, but by most standards, it seemed moderate. Usually it consisted of a couple of beers at Roger's Bar after work, with an occasional lingering that was only a little longer than what was prudent. I only recall seeing him drunk once. Or was it twice? The few images blur together, and I don't think it ever interfered with work or financial obligations. Nor did it translate into any abuse on the home front. Growing up, I never saw a beer in the refrigerator, never saw a mark on Dad's driver's license, never saw him buy any kind of alcoholic drinks from a store, never saw Mom staying up into the wee hours because he didn't come home on time. But Dad's moderation was no great comfort to Mom. She had grown up during the older generation's supposed perpetual supply of "good old days" that were too often only half descriptive of that reality. In the 1930s, people did not talk much about rehabilitating alcoholics, and few people whispered a word about abusive fathers, both of which Mom had to face in her early childhood and thereafter. The corporate misery locked into that dreary Depression decade filled a continent, but most of the groans were not heard. By the time Mom was eighteen, she could no more tolerate a little bit of drinking than a woman can be a little bit pregnant: it was an all-or-nothing proposition.

The thing followed Dad home one night. I knew something was wrong when Joe Girda, an office mate, came in with him. Joe lived many miles in the opposite direction and had not, to my remembrance, ever before set foot inside our door. Dad was unusually boisterous and loud. He insisted on Joe staying, perhaps because drink makes some men overly social, perhaps because he was still lucid enough to want to put off the inevitable domestic storm brewing on the near horizon. Joe, sensing the iciness in the air, demurred.

"Take your coat off before I punch ya in the nose!" Dad was too loud again. Joe offered an embarrassed, patronizing laugh as he slipped out of his jacket for the few obligatory moments of discomfort. When Dad's attentions turned elsewhere for a moment, Joe slipped quietly out the door. He had only come to make sure Dad made it home alright.

Other images from that night remain, but I can't put them into a very good order. There is Dad's hand clumsily but affectionately rubbing my head as he walked past. Even at seven or eight I knew he must be drunk, but as he wasn't being unpleasant, I was hoping that the thing might yet be alright. At some point, we were all sitting on the chaise lounge around the kitchen table when the phone rang. He reached for it, an unanticipated and bad bit of luck for us since Dad never answered the phone at home (largely because the phone was a gadfly for him at work). After slurring an unintelligible response, he squinted up at the phone number on the dial and repeated a number that was nowhere near ours, then hung up. We were torn between the two equally disagreeable possibilities: that the caller would call back, in which case we could get to the phone first but probably not without betraying the lack of composure needed for telling the necessary lie, or that the call would not come back, in which case we could probably assume that the worst had been guessed. The call did not come back.

A tense sense of crisis blanketed the household. Mom, apparently concluding that we had seen enough, put Lloyd and me into the car and headed for the Mapletown theater. It is unfair to critique her sense of social planning at such a moment, but her choice of a movie was probably a bad one. For thirty years thereafter, my memories of *Carousel* were shrouded in moody darkness. It was not the moment to see a story about blind love and an unstable marriage. Each of the magnificent Rodgers and Hammerstein songs, sung brilliantly by Gordon McCray and Shirley Jones, must have stabbed through her like a dagger. (Three decades later, Lezlee and I saw the film at the Ohio Theater in Columbus. I again expected to be powerfully moved, but was instead surprised by the flimsiness of a plot that was little more than a handsome Neanderthal slapping around his mousey wife. Those were the pre-*Fiddler on the Roof* days when producers did not want plots to detract from the pool of rich music, lyrics, and voices that they had gone to so much trouble to get.)

Dad slept in the car that night. I was greatly encouraged to learn that Mom had taken out some blankets to put over his oblivious, scrunched form. He also must have taken some comfort from the act when he awoke, but it was soon apparent that the deed was not a peace offering. I don't know exactly how long Dad was in the deep freeze with Mom—three days sticks in my mind, but I might be confusing it with a biblical reference—but what seemed like an eternity passed without a word passing between them. The few suppers we shared together under those circumstances were pretty grim affairs. Short question and answer exchanges passed between either parent and either boy, but, usually sooner than later, these would hit the inevitable dead ends. The circle had been broken.

A day came when the problem was gone. Mom and Dad were talking; the family was one again. I guess the 1950s was the last lingering era when husbands and wives saw marriage as something to be straightened out, not fled out. Still, it took a remarkable couple to clear such hurdles. Mom told Lloyd and me that someone at the bar had spiked Dad's drink with additional liquors that night, and the situation had gone downhill from there. We were greatly relieved. It doesn't take a lot to put a child's world right-side up again, and we eagerly accepted this answer. Whether Mom did, I can't say. But for us, it meant that the nightmare had been someone else's fault, that everything could be explained away. We were happily satisfied.

It was not quite so easy for Mom. We could have seen it in her eyes as she talked to us, had we wanted to understand such complex things. The spiked drink might have explained the drunkenness, but it didn't explain the drinking. It is tempting to conclude that Mom was just too puritanical about this subject, that she had no right to impose on anyone else, including her husband, the failures of her own father. Along the way there had been other hints of her rigidity. One night after she and Dad had attended a theater production with another couple, she had walked out of the club where they had gone to finish the evening. The circumstances were unusual anyway: the choices of a live play and a nightclub were so far off Mom and Dad's social radar as to suggest what must have been a rare business obligation that Dad could no longer put off—but the little pieces of the tale that filtered through to me indicated that Dad had been embarrassed and a bit piqued. I don't remember if he actually used the word "prudish," but the strong hint was there. Had Dad been the type to take his quarrels outside of his family he could have gotten a great many people to agree with him.

The psychologists and sociologists of the time might have also been sympathetic. Post-War America was effectively disposing of what remained of the "independent man/rugged individualist" view of the American male. That had worked well enough for the Isaac Bacons and David Hudsons in the early Reserve, but my father had faced more complicated and dark forces, many of which were far beyond his capacity to understand, much less control or influence. He had lost his mother at four, hit the Depression at five, was on his own at fifteen, married and in a fifty-million death war at eighteen, had heard from overseas the news that his first child had been stillborn, and at twenty-four saw his young wife need to leave a baby and a two-year-old to go to a mental hospital with all of its then-lifelong implications for emotional instability, financial devastation, and social stigma. Many others, including medical professionals, would have joined Dad's work companions in thinking that he was due that beer or two at Roger's Bar. All the more convincing on that point is the fact that I never heard my Dad use those arguments himself. The inner

Mom and Dad. 1945. Married at 18, then immediately separated for three years—compliments of World War II,—their odds against a good marriage were long. Add to these the fragility of fractured families of their own, and the slim-to-none chances slipped closer to the latter. But theirs was the turn-around generation in the Knowles family, fueled by a determination to make the thing work. And so, we live. And thrive. *Photo from author's collection.*

JUNE 1945 LORRY HOME ON NAVY LEAVE, AFTER 27 MONTHS OUT IN THE PACIFIC SOMEWHERE.

part of him still retained that old-line accountability. If he was going to do something he wasn't very proud of, he at least would not let his sons hear him making excuses for it.

As a little boy, I used to wish that Mom would not get so upset about Dad's drinking because I could not stand to see them mad at each other. Now I can see a woman looking, not backward at who her father had been, but forward to who her sons would be. She, too, had been raised in a splintered home, one broken not only by the death of two brothers but also by some of the few things in life worse than death. Somewhere along the way, amidst the wreckage of her dead brothers, the half-remembered nightmares of electrical shock treatments, and the same crushing weight of the incomprehensible world forces of the 1930s and 1940s, she must have looked into the bright eyes of her two boys and resolved that it would not happen again. She would not give the thing the slightest chance. Nor would she bother herself with debates about whether alcohol was a cause or an effect. In either event, it was an enemy, and even my gentle, peace-loving mother knew that you give no quarter to a real enemy. Over the years, she probably lost most of her arguments to Dad, but this was one she did not dare lose.

The lives of Lloyd and me were not the only ones hanging in the balance. Dad's was, too. Like the aftermath of the American Civil War, World War II littered the nation with alcoholics. Humans were simply not made to endure war well—on either the battle front or the home front. Too many of those in the Greatest Generation continued their sacrifices into their late years where they might end their days sitting rheumy-eyed, wrapped in blankets, and squirreled away in dark corners. That was not to be Dad's journey or ending.

Ultimately, Dad won this brewing battle, just as he won most of the important battles in his life. He hasn't had a drink in many decades, and jumped on the wagon without the fine help offered by AA and other support groups. Still, it is too easy to conclude that everything would have worked out alright anyhow; that Dad's remarkably good personal discipline in the face of adversity would have kept him out of any serious trouble as it had a dozen other times in his difficult life; that the moral fiber of the Knowleses would have prevailed. Perhaps. But sometimes I wonder, as he does, and then the "what ifs" come crowding in. What if Dad had been married to a woman who drank with him, who would unpack the six packs along with the eggs and bread? Or what if he had married someone who dared not question him about such things, who enabled his denial, and who instead helped to smooth away the wrinkles of honest doubt in his heart? What if he had married a woman who simply quit on him? The quiet truth about strong Knowles men is that their greatest strength lies in their rather consistent capacity to marry strong, good women. I suspect that this truth was never more in evidence than when two world-weary eighteen-year-old kids eloped to Napoleon, Ohio, during America's lowest ebb of the Pacific war, and faced one of the world's worst winters: December, 1942.

~~~

The fullness of seasons in the Western Reserve, especially along the valley of the Cuyahoga River, illustrates a fact not likely to be guessed by those who have never been there: it is arguably the most beautiful section of Ohio. To the outsider looking at the yellow, metropolitan clutter with which Cleveland and Akron-Canton blanket a good deal of the river on the map, the statement seems absurd. So, too, for many commuters who never lift their eyes high enough to see over the ocean of asphalt and billboards, and whose paths are limited to the treks between the various mile markers on I-71, I-77, or I-271 that separate their places on the job from their couch potato perches at home. Under skies that are cloudy more often than those of most American metropolitan areas, it is easy to assume the aesthetic worst about the valley. But that is a mistake. Much of the valley is a scenic wonder. The slide presentation at the Cuyahoga Valley National Park's Visitors' Center on Canal Road shows the area as it really is, captured by the appreciative eye of the camera. The scenes combine the haunting simplicity of a Currier and Ives painting with the verdant, three-dimensional richness of the View Masters that awed our child eyes in the 1950s. Clouds of emerald green hills roll over each other down to the tight banks of the Crooked River snaking along the valley floor. Adjust the calendar a bit and the nearby hills have become large cotton puffs being crisscrossed by brightly clad skiers. At another seasonal moment a beaver's face interrupts the vistas long enough to hint at his busy tasks in one of the natural wetlands below. Then, of course, back to the forested hills that have now become a kaleidoscope of splashing, fall colors so magnificent in their riotous songs that you are surprised to find no one singing. If you're still unconvinced, take a journey through Ian Adams's (Jim Roetzel, coauthor) stunning photographic portrait, *Cuyahoga Valley National Park*.

Winter on the Reserve has a bit of magic in it for those who believe in wonders. Some of this is due to the excessive snowfalls. More than a hundred and forty miles separate Cleveland and Columbus, but the former gets over twice as much snow as the latter. Yet, Cleveland itself is

Brandywine Creek cascades. Below Brandywine Falls. The creeks are the rich, flowing capillaries in the heart of the Cuyahoga valley.
Photo by Tom Jones.

at the wimpy end of the real squall lines. Over in Ohio's "Snow Belt," in Ashtabula, Lake, and Geauga Counties, Cleveland's accumulations are easily doubled or tripled as Lake Erie pushes up huge volumes of moisture into the southeastward moving fronts of Canadian air. One small New York town on the far eastern end of the Lake gets 250 inches a year, which translates into an average of two eight-inch snowfalls a week. The magic sometimes gets a bit out of hand. During the winter of "The Blizzard" in 1978, the barometric pressure in Cleveland fell to 28.28, the lowest ever recorded, accompanied by fifty to seventy-mile-per-hour winds and cold that paralyzed the area for three days. I can only guess at what happened in the Snow Belt to the east. That winter and the brutally cold one of the year before were my first two back in Ohio after living down South for ten years. As is the way of childhood memories, I had retained visions of Ohio winters in which the snow was always a foot deep and the temperatures seldom above ten degrees. That was greatly exaggerated, of course, but the greeting given me by the Ohio winters of 1977 and 1978 tended to confirm the flawed memories. The mostly insipid affairs of the ensuing decades in Columbus, with their endless repetition of cool, dismal rains falling on soggy, brown landscapes, have been a keen disappointment. Even in the Reserve, some of the blustery winters have turned shy, and I wonder if some of the globe's observers are right about Ohio becoming a temperate zone. The images of palm trees and deep beaches of white sands on Lake Erie's shoreline scream a discordant note inside my head.

In 1816, that idea would have even been off the charts of imagination. Old-timers looking back later on that year-long winter remembered it as "Eighteen-Hundred-and-Froze-to-Death" and painfully shook their heads at the memory of snows blowing across the Reserve in June. "That year, settlers walked as far as forty miles to get to mills where they could obtain grain. Both people and animals starved. Frosts in June, July and August almost destroyed their food supply, and pioneers were concerned that another "Ice Age" was descending."[21]

Others frankly concluded that the end of the world was at hand. People wandered the roads with faraway looks permanently pressed into their eyes, occasionally stopping to glance up at the faded orange ball hugging the horizon line and seemingly as confused as they were. No doubt the malarial fevers did not make much of an appearance that year, but neither did anything else. Even Christmas.

In fact, Christmas wasn't much of a happening in the Western Reserve during any of the years before the mid-nineteenth century. Those same Connecticut pioneers who built the picturesque New England towns in northeastern Ohio also brought with them a good many Puritan ideas about the unbiblical foolishness of Christmas. While the Catholic and Lutheran Germans in Cincinnati were belting down their Christmases with barrels of beer, the Yankee Calvinists were viewing the scenes with severe disapproval, muttering of popery and heathenism. On Christmas morning, they got their children up and sent them off to school as usual. Not until 1857, by which time the strain of New Englanders in the Reserve was getting watered down with European immigrants, did the state legislature muster the courage to declare Christmas a legal holiday.[22]

In spite of the Puritan ways, the Christmas traditions invaded and secured beachheads in the Reserve. Christmas trees had been seen in Cincinnati as early as 1835, but a Lutheran pastor, Heinrich Schwan, was probably the first to dare to put one up on church property here. The desecration was done in 1851, in of all places, the Reserve's capital, Cleveland. Santa Claus got his start about the same time, largely through the Reverend Clement Moore's 1844 book, *A Visit From St. Nicholas*. If the vanguard from old Connecticut hoped that the fad would soon pass away, they were to be disappointed. Stockings over the mantel piece, the midnight visitor, and especially the presents, were here to stay.[23]

The merchants, as always, were close on the heels of the traditions, and seemed especially fond of the one that encouraged the giving of gifts. They were, I suppose, the pecuniary ancestors of the modern brood that annually brings us decorated shopping malls in September, ten-pound newspapers on Thanksgiving, and an endless stream of images that turn a sixty-minute TV football game into a three and one-half hour orgy of guilt and desire. Department store Santas are looking less like Burl Ives and more like Fagin, with one hand on the child's knee and the other in Dad's pocket. His elves carry cameras instead of workshop tools. The other people in the store are not merely behind you in line, they are pushing you along, like a prison inmate in the chow line, expected to make the right mechanical reactions with your body at the right times. Most of the piped-in Christmas music is being sung by dead people, but that doesn't matter much since everyone knows that most of this is facade anyway. A dead person can pull your Christmas strings as effectively as a live one in Sears.

Cynical? Well, I suppose so, or maybe I'm just trapped at the crossing point of two, old Christmas tensions that went at each other hammer and tongs on the Western Reserve and in the Knowles family. If ever there was a family caught up in the holiday dynamic existing between the New England thesis and the German antithesis, it is ours. No one has more fun *and* more frowns at Christmas than a Knowles. I weave uncertainly along the line separating seasonal richness and raunchiness, shaking my head in wearied disgust but smiling involuntarily at the thought of my granddaughter's face when she opens a gift. No matter how much strain Christmas brings, I find my heart yearning for it when the chilly winds of November blow the last of the stubborn leaves from the trees. My mind skips to memories of backyard football games in the snow, of two boys and their dad searching for a Christmas tree in the late afternoon twilight of a December day, of a Lionel Train running on an oval track bringing a pile of presents that had magically appeared under the Christmas tree when I momentarily lapsed in my Christmas Eve vigil, of church cantatas, and caroling in the institutional light and baking heat of Hawthornden State Hospital (where Dad once started "Hark the Herald Angels Sing" to the tune of "Deck the Halls," and a patient joined us in our hearty chorus of "goodbyes" as he tried to exit with us). I also think of Christmas night at Aunt Rosabel and Uncle Max's in Bedford, where the larger family gathered each year in the big, tumbledown house that for many of us in the tenth generation of American Knowleses was synonymous with Christmas.

In a way, that night belonged to Grandpa Knowles, probably the only one of the year that did. He was the embodiment of that ancient Puritan-German tension, frowning at the mistletoe above the doors and the skirt lengths above the knees, then boisterously joining in the singing of the carols and devoting his best moments to the making of his special punch. I always had my doubts about that punch. No doubt Grandpa thought the world was enriched by his annual concoctions, but I would have preferred him to be more of a Puritan at that point. It seemed like everything within reach that was nonalcoholic ended up in that bowl. The only thing we knew for certain was that it would be red, but even the cherry pop or Kool Aid had to fight to maintain its color against the barrage of other ingredients. One Christmas night he had taken this delightful chore to the basement where he toiled away over a large pot in a sooty corner. Probably the women had gently nudged him out of the kitchen upstairs, and with the press of people and presents in every other room, he believed this was his only workplace. He may have been right in that, but the choice was still a poor one. Uncle Max's basement was one of those typical of early twentieth century homes in which large, dark rooms struggled to house monstrous coal furnaces shooting off arm-like ducts in every conceivable direction. There was no place for cozy paneling or fluorescent lights tucked neatly into drop ceilings. When people went down there, they were just as likely to call it the "cellar."

But on Christmas night, there was an added complication for the Havens' basement. We kids used it for our tricycle races and ping-pong games after our nervous energy had brought the inevitable "G'wan downstairs and play!" from some stuffy adult. It was in this dingy and frenzied environment, illuminated by a solitary bare light bulb suspended from the ceiling, that Grandpa had chosen to make his punch that year. At some point an errant ping-pong ball went zinging momentarily into his line of weakening vision before bouncing harmlessly away. Having neither seen nor heard the ball's escape, Grandpa increased the decibel level of his mutterings and plunged both hands into the punch up to his elbows, sweeping the murky bottom of the punch pot for the intruder. At first I was amused by Grandpa's lapse in the laws of elementary physics, but then, remembering the taste of the stuff, concluded that a ping-pong ball might, indeed, sink in it. I drank milk that night.

Grandpa never did get the hang of gift-giving at Christmas. I suspect that this is true for most people who were adults during the Depression—

the extravagance must seem discomforting, if not downright wasteful—but this would not have been Grandpa's strong suit under any circumstances. He was generally suspicious of any activity that detracted from the religious aspects of the holiday and, besides, gift-giving meant keeping track of too many people. He never could remember who gave him what. If you gave him a tie there was a fair chance that you would get it back gift-wrapped the next year. One memorable Christmas night, he gruffly called Lloyd, Jack, and me together to dispense with some of his valuables. We eyed each other warily as we dutifully gathered round him, having long-since learned that what was valuable to Grandpa might be less so to us. (He had been known to include in his letters samplings of his ear wax.) But this time, Grandpa's gifts succeeded brilliantly—more so, in fact, than he had anticipated. In his hands were three magnificent marbles, obviously antiques, and clearly agates from a time when marble-making had been an art. Two in particular were beautiful beyond description, making our "purie" and "catseye" boulders dull by comparison. Here was some real currency for boys in the 1950s. Lloyd and I had a wastebasket full of marbles at home, the majority of them won during countless school recesses at the Northfield Elementary School, but the haunting gems jostling coyly in Grandpa's unsteady hand would be the king and queen of the lot.

But life is inherently uneven, and, so too, was the match between the number of boys and the number of prize marbles. Grandpa saw the problem as soon as all six hands reached for the same two marbles. His solution was to mix them around roughly in his cupped hands, then arbitrarily hand them to us one at a time. But the capricious fates proved as unjust as any blatant favoritism. Lloyd and I got the two beauties right off the top of the lot, while Jack had to settle for the poorer specimen. Tears rimmed

Jack's eyes, but to his credit, he never verbalized the obvious complaint that might have spoiled Grandpa's gift-giving ceremony. I felt genuinely sorry for Jack that night—almost sorry enough to trade him my marble.

It seems odd that those Christmas nights should have been so important to Grandpa. Like most older people, especially when they find no one else within twenty years of their age, Grandpa was easy to ignore in the swirl of parties. He could be seen on the fringes of the activity centers, walking in his distinctively stooped and hurried gait, as if he had something important to do in some other part of the house. Except for hellos, goodbyes, and his long devotional before we began unwrapping presents (too long for us squirming kids), Grandpa was seldom the center of attention. But somehow, for him, the whole must have been bigger than the sum of the parts. A few weeks before his death in December 1974, at age eighty-seven and almost completely mentally separated from what little remained of his earthly body, Grandpa's cloudy eyes momentarily blinked through the late afternoon grayness of the nursing home as he asked Mom and Dad, "We *are* getting together for Christmas this year, aren't we?"

He, of course, did not make it to the gathering. But, probably out of an old and cherished habit, he clung to his life until December 29. I am sure he is waiting for the next full gathering of the clan.

Notes

1. Ellis, *The Cuyahoga*, 30-31.

2. Ellis, *The Cuyahoga*, 39-43.

3. Grace Goulder Izant, "David Hudson: The Howling Wilderness," In *Ohio's Western Reserve: A Regional Reader*, edited by Harry F. Lupold and Gladys Haddad (Kent: The Kent State University Press, 1988), 69.

4. Harlan Hatcher, *The Western Reserve*, Revised Edition (Cleveland and New York: The World Publishing Company, 1966), 52-53.

5. Jackson, *Colorful Era*, 9. The bridge, by the way, is the same one much in the news in 2012 when bomb plotters were foiled in an attempt to take it down.

6. Ellis, *The Cuyahoga*, 67.

7. Gertrude Van Rensselaer Wickham, ed., *Memorial to the Pioneer Women of the Western Reserve* (Cleveland: The Woman's Department of the Cleveland Centennial Commission, 1896), 432-438.

8. Charles Whittlesey, *Ancient Earth Forts of the Cuyahoga Valley, Ohio* (Cleveland: Fairbanks, Benedict & Co., Printers, 1871), 12.

9. William Henry Perrin, ed., *History of Summit County* (Chicago: Gaskin & Battey, Historical Publishers, 1881), 571.

10. Bessie Gooseman, *A History of Olde North field Township* (Northfield, Ohio: Historical Society of Olde Northfield, 1973), 27.

11. Perrin, *Summit County*, 574.

12. Bliss, "Pioneer Women," 434.

13. Perrin, *Summit County*, 573.

14. Condon, *Cleveland*, 22.

15. Ellis, *The Cuyahoga*, 53-54.

16. Izant, "David Hudson," 68.

17. Perrin, *Summit County*, 571.

18. Perrin, *Summit County*, 568.

19. Crary, "Frontier Living," 73.

20. Crary, "Frontier Living," 74-75.

21. Lar Hothem, "Link With a Lost World," In *The Western Reserve Story*, edited by Will Folger, Mary Folger and Harry Lupold (Garrettsville, Ohio: Western Reserve Magazine, 1981) 112.

22. Ann Natalie Hansen, *Westward the Winds* (Columbus: Sign of the Cock, 1974), 112.

23. Hansen, *Westward the Winds*, 112-113.

8
Which Way is Up?
The Faith

Church and state. Hale Farm. Religious expression was a varied and forceful backdrop in the Western Reserve. America's strongest indigenous church movements were, in important ways, molded here. *Photo by Ian Adams.*

Mormons, Shakers, Disciples, and other off-the-main-drag religious groups that thrived in the early Western Reserve • Mind v. emotion: the battle for the biggest piece of God • Asking the only important questions

M yths abound in the Valley of the Cuyahoga. There is one that says that religion came to the valley in the form of a sturdy Protestantism brought by Connecticut Yankees who were either seeking greater freedom for their faith or, in pursuing the call of the frontier, brought with them the solid faith of their fathers. The myth comes apart at two points: The first Connecticut landowners in the Western Reserve were neither the first religious people there nor a particularly religious lot. Long before Christ, or Moses for that matter, people were worshiping their Creator. They filled their mounds with articles for the afterlife, revered their priestly shamans, and ceaselessly wondered at the divine force the Algonquians variously called the "Mystery of Light," "Great Spirit," "Supreme Being," "Infinite One," and "The Creator." Their seemingly primitive capacity to see God in every particle of nature rings closer to the truth than our secular perspective that sees Him in none of it. And, while they also knew of the Evil Spirit, one modern Lenape wag concluded that they "never knew the meaning of the word 'HELL' until after the white man came to America!"[1]

The same white men who brought too much of hell were also guilty of bringing too little of heaven. In most cases they were running from, not to, something. Even the pious David Hudson appears to have been at least partly motivated by his desire "to remove myself to the solitary wilds of the Connecticut Western Reserve where my former sins were unknown."[2] Those less troubled by such pricks of conscience were as likely to be motivated westward by Connecticut's played-out soil and increasingly rough winters as by any desires to proselytize the natives or enjoy religious freedom.[3] Like their forebearers who journeyed to a new land, for every boat of Puritans who landed, there were uncounted other boats unloading debtors, convicts, charlatans, and other assorted adventurers.

The Connecticut Yankees cannot even make the claim to have been the first white religious influence in the Reserve. Over a century earlier, the irrepressible and fearless Jesuits had braved northeastern Ohio, preaching their message to the wandering, mysterious tribe that tended the valley prior to the Iroquois invasions of the seventeenth century. There were other Christian forerunners as well. When Moses Cleaveland and the other directors of the Connecticut Land Company fixed their covetous eyes on the early Reserve, they were often looking at the maps of John Heckewelder, the Moravian missionary to the Delawares who succeeded in converting a large group of them. It was Heckewelder and his Moravian compatriot, David Zeisberger, who first ventured into the Cuyahoga Valley as pilgrims in search of religious security. Cruelly treated by both Indians and whites (the massacre of the defenseless Moravian Delawares at Gnadenhutten by American colonials ranks as one of the worst episodes in the sorry saga of American-Indian affairs), the Moravians wearily settled in for a winter on the Cuyahoga at Tinker's Creek. Since there had as yet been no Joseph Tinker to lend his name to the stream, the Moravians called the place Pilgerruh.

In fact, the vast majority of the nation's frontier people—Church of Christ historian Leroy Garrett states that the number is in excess of ninety percent—aligned themselves with no religion at all. The period separating the two wars with Britain saw what still stands as the nadir of the nation's religious fervor west of the Allegheny Mountains. That was the same period in which the youthful David Hudson allowed himself to be caught up in the earthy vagaries of French Revolutionary thought, and the Western Reserve was being settled.[4] Joseph Badger, a tough-minded, circuit-riding Calvinist who had been further steeled by service in the Revolutionary Army, found the pickings extremely lean when he began his lonely missionary journeys through the Reserve around the turn of the nineteenth century. While he managed to get a church planted in Hudson as early as 1801, Badger probably received warmer receptions among the Indians than the whites, the latter of whom "had been through all of that" back east. He listened disapprovingly as Ravenna founder, Ben Tappan, made a July 4 speech that was "interlarded with many grossly illiberal remarks against Christians and Christianity." Freely mixing patriotic and anti-Christian remarks only hinted at the size of Badger's problem. Some of those that did bother to show up to hear his preaching in some friendly cabin were often malcontents for whom any break in their lonely isolation was a welcome relief.[5]

Badger was lucky if half of his audience was neither drunk nor hung over. A large number of our pioneer predecessors were alcoholics, pure and simple. Whiskey was the one thing never in short supply among them. Surveying the brutally tough lot of the early settlers in Boston and Peninsula (the latter settlement boasted fourteen bars), Randolf Bergdorf hints that a booze problem seemed a natural consequence of a very tough existence. "Dwellings were crude, tree stumps hindered farming, bartering was prevalent, roads were negligible, disease was rampant, farms were isolated, and whiskey flowed like water."[6]

After reviewing all aspects of pioneer life in Ohio, Ann Natalie Hansen came to a similar conclusion: "The most common scourge of the frontier was drunkenness; astronomical quantities of hard liquor were produced and consumed."[7] Ministers like Joseph Badger were expected to take their grog like everyone else. The myth of the "religious pioneer" did not begin with those missionaries. They came to reach sinners, not welcome saints.

~~~

From the beginning, the land of the Cuyahoga was a magnet for bizarre religious thoughts and groups. After the fanatical Jesuits and ill-fitting Moravians, Appleseed John Chapman came to the southern fringes of the Reserve, planting trees, poking needles into his flesh, gathering curious herbs, buying mistreated animals solely for humanitarian reasons, and otherwise spreading so much love and kindness in his wake that the Indians could not bring themselves to harm him. Chapman's ragged clothes and trademark cooking pan belied his education at Harvard, then primarily a religious school, and hinted at the cultic Swedenborgianism that formed his religious views.[8] During his thirty-seven years of orchard planting in Richland County, Chapman saw other oddballs also take a liking to the wild religious airs of the Western Reserve, including Mormons, Shakers, and Disciples of Christ. Mainline denominationalists like Joseph Badger shook their heads in disgust, but others were also growing concerned. An editor of the *Painesville Telegraph* voiced the growing complaint in April 1835:

> No other country on earth can boast of such varied forms of religious sects and such palpable departures from primitive simplicity and purity of the Gospel, as this country. We would not forge chains nor bind fetters around any human mind, but we would gladly see public sentiment frown upon those mental hallucinations that disgrace Christian lands, and shun communion with those preposterous forms of worship that are merely solemn mockeries of Religion![9]

The Cuyahoga region, as well as most of America's then-great northwest, was ripe for a spiritual explosion at the beginning of the nineteenth century. Nature abhors a vacuum, spiritual as well as physical. Sooner or later, the very rocks will cry out if the need is not being met. At the dawn of the Christian faith, in the well-regulated but spiritually starving world of the first century, the Gospel spread like a prairie fire against the backdrop of the silly Roman gods and brutal repression. Circumcised men with a thousand-year-old habit of keeping to themselves were suddenly fanning the flames of a new mystery religion into the far corners of the known world. So, too, did the Second Great Awakening of the early 1800s sweep into pioneer lives made of misery, loneliness, alcoholism, fatigue, and despair. The people on the frontier responded with a throbbing surge toward the new/old Gospel message. Not surprisingly, some of this pent-up power vented off in some rather odd directions.

In many ways, some of the newer religious movements that so troubled the *Painesville Telegraph* editor were similar to the cults that would come to the nation's awareness in spectacular style in the latter twentieth century. They were often characterized by reliance on a charismatic leader, communal living, common ownership of goods, faith healing, heavenly visions, a heavy impressionistic impact on the young, and persistent issues relating to sex and money. The response to them from the Reserve's populace was frequently violent, and consistently frosty. Some of the area's bleaker historical moments center on the treatment of Ann Lee's Shakers and Joseph Smith's Mormons.

Like many names born from notoriety, the name "Shaker" was affixed from the outside and for purposes of derision. The followers of Mother Ann Lee of England called themselves the United Society of Believers in Christ's Second Appearing, or the Millennium Church. The derogatory term, shaker, reflected the shaking motions that palsied the faithful during worship, but which, by the mid-nineteenth century, had evolved into a graceful dance. Today, it is hard to envision any group of people less reflective of Shaker Heights, one of the two or three richest communities in the United States for many years. The rows of castles that line Shaker Heights are open volumes on the excesses of private enterprise, ownership and other marks of "getting." In stark contrast, Mother Ann's humble followers devoted their lives to giving and giving up—their homes, their labors, and most notably, their sex lives. They were disliked for their strange ideas about celibacy and communalism, but redeemed themselves in the public eye with a solid work ethic, economic intercourse with their apostate neighbors, and a genuinely humble spirit. Ultimately, Cleveland area folks seemed to conclude that if they left them alone with their funny ways and self-defeating ideas about procreation, the Shakers would just fade away some day. And so it happened.

The Mormons evoked a less tolerant response. Known today as The Church of Jesus Christ of Latter Day Saints, or simply the LDS church, this was a people who fed on the daily visions of Joseph Smith, unlike the Shakers who had been bereft of Mother Ann Lee's presence since before their Cuyahoga settlement days. They carried guns when they had to, and even opened their own money-printing bank in an effort to put distance between themselves and the none-too-friendly infidels who surrounded their home in Kirtland, Ohio. Mormonism is a combination of the fascinating and the fantastic. They believe that a remnant of the Israelites from Moses' time found their way to America, and that Christ ("a tall white man, bearded and with blue eyes [who] wore loose, flowing robes.") brought his ministry to this continent after his resurrection in Palestine.[10] Essentially, the entire core of the faith rests on Joseph Smith's finding and interpreting the golden plates that became the Book of Mormon, the proofs for which are contained in signed statements of witnesses. By Smith's account, the tablets were, for unknown reasons, written in a curious combination of Egyptian, Chaldaic, (Babylonian), Assyriac, and Arabic, but divine guidance led him to translate them despite a lack of formal education that left even his English proficiency in doubt. Too, his translations, as well as his subsequent visions, always came out in King James English.[11] All of this happened while Smith was still a young man in New York, but nothing much came of it until he and his scattering of followers came to Kirtland near the headwaters of the Cuyahoga.

Not surprisingly, the Mormons took a sour view of their Shaker neighbors living near the eastern rim of the cup cut by the Cuyahoga. A faith living on the daily visions of one man cannot afford to be too generous with the visions of others, and visions were in no short supply among the Shakers during Smith's brief time in Ohio. The Prophet was also equal to that task, dutifully passing along words from God just as he had received them: "Hearken unto my word, my servants Sidney, and Parley, and Leman; for behold, verily I say unto you that I give unto you a commandment that you shall go and preach my gospel which ye have received, even as ye have received it, unto the Shakers."

Neither did the vision scrimp on the particulars of the Shaker beliefs. With a disapproving nod in the direction of Ann Lee, the Lord emphasized that "the Son of Man cometh not in the form of a woman," and, furthermore, God had never said anything about celibacy.[12]

It was natural that Smith's visions drew more skepticism from competing religious interests than from the ever present group of "infidels" that formed in cynical clusters at the edges of early nineteenth century public worship services. The latter had only to satisfy some curiosity or stereotype; the former had to satisfy their spiritual vested interests. There is no shortage of stories concerning the credibility of Smith's visions. Disciples of Christ historian Henry K. Shaw tells a story of two Disciples, J.J. Moss and Isaac Moore, who feigned an interest in Mormonism to investigate the claim that Mormon nighttime baptisms in Kirtland were accompanied by visions of angels walking on water. Upon closer investigation in daylight, the two men found a wooden plank extending out into the water just below the surface. Moss and Moore apparently concluded that such a prop was an insult to any self-respecting angel, whereupon they sawed half way through the boards then retired. That night both the believer and the angel were baptized. In another anecdote, Shaw says a man named Symonds Ryder, called to the Mormon eldership during one of Smith's visions, refused the honor when the Lord misspelled both his first and last names.[13]

But the Latter Day Saints were not to be so lightly brushed aside. They seemed to thrive, not in spite of, but because of, the persecution that they consistently attracted. Ryder, a Disciples minister who was probably embarrassed by his brief infatuation with Mormonism, did not completely assuage his anger with his rebuff to Smith's offer of eldership. During a visit to Hiram, where Smith and a fellow Mormon leader had made great inroads, Ryder led a mob against the two during the night. As always, religious violence seems to carry its particular degree of ugliness and cruelty.

> Smith was asleep in a trundle bed when the mob grabbed him. They tore off his clothes, beat him, and tore his skin. They cut his lips with a glass vial that broke when they tried to force it into his mouth. Then they smeared his body with tar and covered him with feathers from a pillow that they had ripped open. They dragged both of them out into the cold night, bumping them over the frozen ground, and finally left them unconscious in the field.

Smith's wife fainted when he crawled back to the house. Several women spent the night scraping away the hotly splashed tar now frozen to his body. It was a deeply despairing moment in his life, but the remarkable man determined that he would sink no further. The next morning he astounded several of the mob members, who had shown up at his church to see what amusing developments might transpire, when he mounted the Sabbath pulpit and preached "quietly and eloquently, as if nothing untoward had occurred."[14]

Smith could not be beaten out of Ohio, but money troubles were a different matter. The dark winds that swept the young nation into economic panic in 1837 caught the Mormons at a bad time. For various reasons, the Mormons never could match the Shakers' success in the business world of the Western Reserve. This, and the ever-growing flow of indigents attracted to his communally owned properties in Kirtland, forced Smith deep into dept in order to keep the work afloat. Smith, himself, and a couple of his closest assistants owned several significant pieces of property that had been deeded to them by the faithful. When it became clear that the problem of indebtedness was not likely to go away, he did what everyone else was doing in the Reserve in the early 1830s: he started his own bank. The state already had forty-two, nine of which were unchartered, and the state legislature was being petitioned for nearly as many more. And the Mormons were rich in the second-best capital in the west, land. When the legislature denied the Mormon charter request, the Prophet assumed, not unreasonably, that this was just more persecution. He concluded that while the state had refused his bank charter they had said nothing about flying under slightly different colors. Hence, the $4,000,000 in notes they had printed for The Kirtland Safety Society Bank were reprinted—apparently without a blush—as the notes of the "Kirtland Safety Society Anti-Banking Company."

Times being what they were, the ploy worked for a while. Westerners were used to odd currencies. It was during the same time that James Brown was busy down in Boston Township deliberately and successfully passing off his counterfeit money as counterfeit, such was public confidence in the wide acceptance of his illegal tender. And, at any rate, there was all that Mormon real estate behind the notes, not to mention the prosperous image of the great Temple that the Mormons had constructed by each donating one day of labor a week. (The Mormon women had donated their greatest treasure, the glassware and china that had been shepherded with such painstaking care over the mountains and lakes, all of which was smashed and added to the stucco for the temple's exterior walls.)[15] Too, when some people began to worry about what kind of specie stood behind the venture, Smith obligingly opened the vaults to exhibit rows of boxes topped with silver coins and signs proclaiming the one-thousand-dollar contents of each. Under the thin veneers of coins were hidden sand, lead, scrap iron, and anything else that would approximate the right weight. But the New York banks balked at the Mormon notes from the beginning, and the first hint of actual suspicion brought the whole sordid episode down around the Prophet's head. He was arrested repeatedly, bailed out just as frequently by his hard-pressed followers, and saw many of his own people turn bitter against him, including six of his twelve apostles of the church. He and his faithful companion escaped the Cuyahoga valley's claws for good one January night in 1838, barely ahead of an armed mob that Smith claimed continued their enraged pursuit of him for two hundred miles.[16]

Eventually, the "mobocracy" that Smith engendered wherever he went cost him his life. In 1844, while offering himself to the nation as a presidential candidate on the same ticket with the Mormon brother who had shared his tar and feathers in Hiram, Smith was murdered under a hail of bullets while sitting in a Carthage, Illinois, jail cell. But the Mormon faith would not die. Smith's vice presidential running mate and close advisor, after failing to secure leadership of the group, returned east with a Mormon remnant to live out his years in bitterness and confusion, but

Brigham Young took the main group west to Utah and a safe haven for the vibrant faith that pulses through the veins of the Latter Day Saints Church, now stronger than ever.

∼∼∼

When the Mormons finally took leave of Ohio, they left behind their great temple, a trail for the many thousands who would follow westward, and an uneven religious legacy that caught at the fabric of the Knowles family faith. Chief among these was Sidney Rigdon, the Prophet's heretofore unnamed lieutenant, the man who had shared with him Hiram's horrible night, that frenzied midnight flight out of the Western Reserve, and the 1844 presidential ticket. But long before Rigdon cast his lot with the Latter Day Saints, he had been a powerfully effective minister for a third group inhaling the unique religious winds blowing in the Reserve. A good many Knowleses of bygone days would bristle at my mention of the Disciples of Christ in the same breath with Mormons and Shakers. (In fact, the linking *is* a bit misleading except for the Rigdon connection and the fact that the Disciples did attract some suspicion and ill treatment, though nothing compared to that suffered by the LDS Church.) But to the mainline Protestants of the early nineteenth century, the Disciples of Christ was just another misguided group rebelling against the authority of the Church. Even today, the three branches of the great nineteenth century Restoration Movement (the name of the historically broad roof covering the several subsequent splits among the Disciples) occasionally gets waved off as a sect or cult.

Some background faith history of the Restoration churches is in order for anyone who cares to understand the nineteenth century Cuyahoga valley, in general, and my extended family, in particular. Too many historians, particularly in the twentieth century, tended to neglect religion as a motivating factor in human behavior. Political, economic and social forces are more easily tracked in the trail of "hard" information left in their empirical wakes. Measuring the motives of human souls requires harder work. But the evidence is there—in the letters, newspapers, diaries, and chambers of government from a time when key American leaders were not afraid to discuss religion in polite society. We are, after all, highly spiritual creatures.

The hunger of the Second Great Awakening, which opened hearts to the primitive power of the Gospel, also opened minds to better ways of pursuing it. The last thing in the world the Campbells, son Alexander and father Thomas, wanted to do was start a new church. They spent many frustrating years clinging in the increasingly chilly atmosphere of the Presbyterian and later Baptist Churches before being forcibly cast off. Ironically, the passionate plea emanating from this contentious group in their painful pursuit of New Testament Christianity was one for Christian unity. The brutal irony in this Restoration Movement, as it came to be known, is that what began as a quest for unity among God's people ended up creating another church, one in fact that became the largest religious movement indigenous to the United States with the exception of the Mormons. Variously called Disciples, Christians, and the less friendly term, "Campbellites," the heirs of this movement today include more than four million people.[17] Along the way, they also included many Knowles ministers and numerous family members spiritually teased by the prospect that religion could be as simple as "No book but the Bible; no creed but Christ."

It was, in fact, the Presbyterians' reliance on creedal practices that first drew the critical attention of the Campbells. Alexander's limit was reached before he turned twenty-one when he found that he could no longer tolerate the practice of church leaders severely scrutinizing members in order to determine their fitness for taking the Lord's Supper. The process even included the literal awarding of a token to verify the supplicant's worthiness for the exclusive ceremony. The young Campbell could find nothing in scripture to justify this and a host of other practices, nor could he find in the church some other practices clearly called for in scripture. Like the Catholic Church of Luther's time, the Protestant denominations seemed to be devoting much time to the divisive sectarianism that inevitably follows the rules of men, and virtually no time to the unity promised by devotion to the words of scripture. Campbell, still in Scotland at the moment when he discarded his unredeemed token and walked out of the communion service, did not yet know that his father, who had preceded his family to America, had come to the same kind of conclusion at the very same time. But the Campbells were not the first to make the discovery that a return to New Testament Christianity was needed. The idea seemed to be occurring to several others, independently, at about the same time. Men like the Haldanes in Scotland and Barton W. Stone in North Carolina (later Kentucky) had been struggling with the issue for years by the time Alexander Campbell entered the fray.

It is not surprising that the idea found its most fertile ground in the American west of two centuries ago. There was more suffering on the frontier, and where there is more suffering, there is a more strenuous seeking of God. Descriptions of this spiritual hunger are virtually incomprehensible to modern Americans. They tell of people traveling for days in springless wagons over severely rutted roads to sit on backless wooden planks for long hours under hot suns over many days listening to Campbell debate men like Utopianist Robert Owen. Campbell was a lively debater, but Owen spent his part of the nine day event reading and rereading—twelve times, in all—his twelve laws of human nature, yet still they listened. They tell of Campbell's rag-tag little paper, *The Christian Baptist*, published out of his back pocket, sweeping across the vast western lands like a modern movie box office sensation, its words slaking the thirst of dried souls. They tell of men like the great preacher, Walter Scott, an unknown and solitary figure riding quietly into town in the morning, intriguing some of the idle children with his five fingers of faith exercise, perhaps giving a few cents to some of the older ones to announce his presence door-to-door, then returning in the evening to a schoolhouse jammed with people desperately interested in hearing this man who had put such remarkable words into the mouths of their children. They tell of the indefatigable drive of men like Campbell and Rigdon who, after experiencing arduous days of hours-long preaching and enduring hostile and vulgar attacks, would sit up all night in feverish conversation about this consuming passion in their lives. On the East Coast, people were beginning to enjoy more comfortable spiritual surroundings, but in places like the Cuyahoga Valley pain was forcing people to remember who they were and what was ultimately important in life.

Campbell's status as one of the nation's preeminent men of the first half of the Nineteenth Century also sets him apart from religious leaders two centuries later. All of our great ministers today are tucked away somewhere in the nation's mental pigeonhole that is reserved for "religion," which, after all, is not nearly so important as those reserved for politics, employment, sports, and the other roles played on our societal stage. Here in Columbus, our local paper easily slipped into the journalistic fad of limiting the "religion page" to Fridays. But such distinctions were nowhere to be found in Campbell's world. Ministers were part of the cultural warp and woof. Indeed, Campbell qualified as a Renaissance Man, appreciated as a scholar, philosopher, farmer, orator, statesman, administrator, minister, and patriarch. During an 1858 visit to Louisville, the city's newspaper, the *Journal*, honored his presence with a glowing editorial of praise. After formally observing the journalistic tradition of brushing aside his religious tenets ("with which, of course, we have nothing to do") the editor demonstrated not the least hesitation in recognizing Campbell's greatness. "[He] claims, by virtue of his intrinsic qualities, as manifest in his achievements, a place among the very foremost spirits of the age. His energy, self-reliance, and self-fidelity, if we may use the expression, are of the stamp that belongs only to the world's first leaders in thought and action. "[18]

Campbell's circle of acquaintances and admirers included men like John Brown, James Buchanan, Andrew Jackson, James Monroe, John Marshall, John Randolph, James Madison, and James A. Garfield. Madison and Randolf, who shared Campbell's duties as delegates to Virginia's 1829-30 Constitutional Convention, whether agreeing with his antislavery sentiments or not, would leave the convention floor each day to seek out his preaching in various Richmond churches in the evening where Madison found him to be "the ablest and most original expounder of the Scriptures I have ever heard."[19] Henry Clay leaned upon his friend for support in the Senator's monumental struggles to find compromises that would keep the volatile American union intact in 1850.

If any one person possessed the personal influence to nudge the nation off its destructive path during those hate-filled years, it was Alexander Campbell. Jefferson Davis thought highly of him, and continued to have a keen interest in Campbell's Bethany College even after Davis's nephew died there as a result of a fall on the winter ice. Lincoln apparently never met Campbell, but he was no stranger to the Restoration preachers in Illinois, largely because his father, Tom Lincoln, had been caught up in that movement. The inflamed Congress sought out Campbell on a late year's eve before the War for a special, informal session convened solely to hear him preach. Names that would be inked into Civil War history books—Daniel Webster, William Seward, Salmon Chase, Stephen Douglas, Thaddeus Stevens, and Clay—sat before him spellbound, entranced by his message that assured them that, "A God who would give His Son because He loved the world would not abandon it now." Similar invitations came from the state legislatures in Missouri and Indiana, the latter attending one of his Indianapolis lectures as a body.[20] In his spare time, he physically built and administered a college, fathered two large broods of children by two different wives (all of his children by his first wife, Margaret, as well as she herself, were dead at least a decade and a half before his final call in 1866), and ran one of the largest and most profitable farms in Virginia; his sheep flock was the largest in the state.

Campbell's close association with Sidney Rigdon is one of the more interesting and significant religious stories of the Western Reserve. On a larger scale, it directly affected a good-sized piece of American history, especially the chapter relating to the Mormon's mark on the American (far) West. It is not stretching the imagination beyond credulity to envision a scenario which, slightly altered from the actual script, might have produced neither Mormons nor Disciples as lasting religious entities on the American religious scene. The fateful moment came during the meeting of the Baptist's Mahoning Association in 1827. Within a few years, even that renegade association would not have room for the reformers whose Restoration plea (i.e., Christian unity via restoration of the First Century Church) had put them on a collision course with other Baptists, but at the time the group was deciding to take the unprecedented step of sending a full-time evangelist into the wilds of the Western Reserve. Walter Scott, who had not planned to attend the association meeting that year until yielding to Campbell's persuasive invitation, was picked for the job. The more obvious choice for that honor was Sidney Rigdon, who had already been ministering effectively on the Reserve for several years and was, in Campbell's words, "the great orator of the Mahoning Association." In passing over Rigdon to get the lesser-known Scott, the association chose the man who would become the greatest evangelist in Disciples' history. Scott averaged an incredible one thousand baptisms a year for thirty years.[21] Among those caught in the widening ripples of his many conversions was James A. Garfield, via Jonas Hartzel and W. A. Lillie.[22] Disciples' historians W.E. Garrison and A.T. DeGroot have concluded that were it not for Scott's efforts, the Disciples probably would not have a history.[23]

Rigdon appears to have a similar claim on the development of Mormonism, and the larger contribution of the Disciples to Mormonism in Ohio cannot be overlooked. It was a Disciples' minister, Parley Pratt—who along with Leman Copley and Rigdon found himself on a first name basis with the Lord in Joseph Smith's revelation aimed at the Shakers—who effected Rigdon's critical conversion to Mormonism.

> Rigdon was their [the Mormons'] only man with any influence at the outset, the only one with a church and a following. Until Rigdon entered the scene Joseph Smith had but six

followers. It was Rigdon's Disciple church in Kirtland, Ohio, that he [Rigdon] had prepared for something fantastic, that afforded Smith his first mass conversions. Rigdon's popularity in northern Ohio enabled him to reach other Disciples, especially in Hiram.[24]

Others imply an even greater, though more sinister, Mormon role for Rigdon. By the time of the First World War, Rigdon's name was being tied to a theory that the Book of Mormon was a thinly veiled plagiarism of a deliberately fictionalized account of the ancient Hebrews coming to America to become what history knows as the American Indians. A man named Spaulding had concocted the work, writing in the style of biblical history, but his manuscript had been thereafter stolen. No trace of it ever surfaced except, said some who had heard the author reading from his original work, in the same distinct style of the writing Joseph Smith claimed to have found on those gold plates. Further suspicion centered on Rigdon. Even some of those not purporting the Spaulding Theory have suggested that Rigdon, not Smith, must have written the Mormon works since the latter was so poorly educated (although such a criticism exposes a double edge to those who also believe that fishermen could write a Gospel). What cannot be denied is that Rigdon was critical to the success of Mormonism at a time when that movement was extremely vulnerable to extinction. Had the emotionally volatile Rigdon been chosen as the evangelist for the Mahoning Association in 1827, Joseph Smith and Brigham Young might be unknown names today.[25]

The causes of Rigdon's radical departure from Restoration principles are not known, at least not in their proper proportions. It is clear that he had built into his congregation at Kirtland an expectation of some great event, and it might be surmised that he saw more potential for triggering such an event in the charismatic Smith than in the scholarly Campbell. Too, both his ideas and ambition had been occasionally frustrated among the several huge Disciples' personalities, a good example being his desire to practice common ownership of property.[26] That the practice turned up among the Mormons in the 1830s is hardly surprising.

Campbell found Rigdon's apostasy especially repugnant. To a man basing the entirety of his position on the authority of scripture, the idea of superseding that scripture with the writings and visions of one man could be nothing less than sacrilege. Yet he also grieved this loss of his friend and spiritual brother. The two were probably closer than were Campbell and Scott, between whom there seemed to exist a certain degree of rivalry. Rigdon had been completely captured by Campbell's brilliance and persuasiveness the first time they met at Bethany, and it was Rigdon who had shared the ten day, three hundred-mile ride with Campbell to the latter's debate with W.L. Maccalla in Washington, Kentucky. It was also Rigdon, in combination with Adamson Bentley, who gave Campbell entry to Ohio and the Western Reserve.

~~~

There is no evidence that Alexander Campbell ever preached in Northfield, even though his voice was heard in Hudson, Aurora, Ravenna, and several other nearby townships. The Free Will Baptists apparently beat him to the punch in the four jurisdictions comprising my home town, holding a rousing five-week revival in Macedonia that resulted in sixty-five conversions.[27] But there is little doubt that the great Disciple had something of a religious experience in Northfield one wretched night in 1836, even if he didn't know exactly where he was. The harrowing journey up the valley reminded him that, "We are never fully sensible how much we owe to the Eye 'which slumbers not nor sleeps,' for our deliverance from harm and from danger," and that nights like this one "more forcibly remind us of our obligation to gratefully entrust our lives to God." More specifically,

> We spent the whole night on the road from Hudson to Bedford, a distance of only twelve miles; thus carrying the mail [by stage] at the rapidity of one mile and three quarters per hour! We had only to walk some four or five miles through mud and swamps, and to abandon the coach some six or seven times to prevent upsetting and the breaking of our bones during the night watches.[28]

Neither do I know which of the roads through Northfield contributed to Campbell's stagecoach grief. Perhaps it was the road from Little York through Northfield Center (Old Rt. 8) or, perhaps, the wagon road Bessy Gooseman identifies coming from Brandywine and passing northwest of the intersection of Highland and Brandywine Roads. If the latter, then Campbell passed over Miller's Hill just east of the former Conrail Line (now the bike trail), scene of our family sled-riding frolics during our growing up years. Walls's Pond was the only hint of a swamp left in the 1950s, but the area's reputation for the muck and mire was well known one hundred and fifty years ago. Writing at mid-nineteenth century, historian L.V. Bierce noted that the road from Hudson to Newburgh, which is northwest of Bedford, was a virtually unbroken bed of clay and mud, and that the stretch between Campbell's starting point that night and the Bacon farm in northwest Northfield was known as "mosquito swamp." "Teamsters," Bierce added, "used to say there were but two mudholes [on the road], and Tinker's Creek bridge separated them."[29]

Had the illustrious Campbell stopped in Northfield there would have been no Knowleses to hear him speak. They were still up in Nova Scotia, and had been since 1761 when they left the New England ground to which Henry Knowles had immigrated in 1635. I guess our family was destined to sit out America's first three wars. Frankly, I'm selfishly rather glad they missed the last one a quarter of a century after Campbell's harrowing night ride through Northfield, or this book might have been consigned to a parallel universe. So, too, for Ailene Miller Williams, whose family history captures a sense of their Canadian irritation with the American Civil War. "This war makes everything dull here. It is nothing but talking about War," says a letter from my great-great Uncle William Ingram Knowles, writing from Maine to the family back in Nova Scotia.[30]

I can't say for sure if the family disapproved of the growing rebellion against Britain—land needs probably trumped politics in driving their move to Canada in 1761—but they never had such qualms when it came to taking on the established Church, an inclination that made them kindred spirits with Campbell. The Knowles tradition of resisting the institutionalized Church started early and lingers still. I can feel it in my marrow every time I flirt with the conclusion that there is little hope for the ornate, self-aggrandizing institutions which claim to represent the New Testament Church. In my worst moments, I wonder if there is nothing left to do but knock the thing down and start over again. Henry

Knowles, who brought the family name to America in 1635, must have had similar thoughts. Having little taste for Puritan Massachusetts, Henry advantaged himself of Roger Williams's liberal tolerance in Rhode Island, being listed as a freeholder there in 1638. He married into the family of Robert Potter, a radical even by Williams's lenient standards.. Potter was one of the founders of Warwick, Rhode Island, and a disciple of the notorious Samuel Gorton, but the Gortonists' first attempts to settle that town got them tangled up with jurisdictional claims of both the Indians and the Massachusetts Bay Colony. The authorities in Boston sent Captain Cook and the militia to arrest the glorified squatters and break up the settlement (which they did quite efficiently, destroying properties and leaving Mrs. Potter dead of exposure in the wake of the assault). Reflecting the unique theology of seventeenth century Massachusetts, the Boston officials saw my eight-times removed grandfather and his Gortonist ilk as guilty of blasphemy and heresy. They saw any challenge to their authority as rebellion against God. Gorton, Potter and a few fellow rebels were put on trial for their lives. They escaped the capital sentence by only two votes, were forced to labor in chains throughout the winter of 1642-43, and banished thereafter under the threat of instant death should they ever return. But the most interesting part of the court's sentence, at least in terms of later family developments, was that during their time of convict labor they were prohibited on pain of death from speaking to anyone other than Massachusetts authorities lest they contaminate the colony's more impressionable Christians with their heresies, which denied "the authority of the clergy and the importance of outward forms of religion."[31] Three hundred and fifty-plus years later it's still hard to find a Knowles who cares over much about the authority of clergy or outward forms of religion.

Somehow, it seems appropriate that this first known act of religious defiance and courage came into the bloodline from the female side of the family (i.e., from Henry's association with his wife's family). Four generations later, another woman married into the family and brought with her the same convictions that Campbell, Scott, and Stone were hammering into the cornerstones of the Disciples of Christ at the very same time. "Married into the family" is, of course, a sloppy phrase genealogically. Except for the name, Lydia Woodworth, mother of my great-grandfather, Thomas Benjamin (T.B.) Knowles, is as much my forebear as is her husband. What's more, she brought into the line an august lineage far more glittering than that of the Knowleses—the Woodworths even managed to get themselves stitched into the forty-seventh scene of the Bayeux Tapestry—which gave early hints of a stubborn individualism. About the time that our "Henry the First" was settling into old age in colonial Rhode Island, one of Lydia's ancestral uncles was fined one pound for playing cards.[32] Great-great-Grandma Lydia took her stand at an even earlier age than Campbell did, and at a time (1816) when the twenty-eight-year old Campbell was yet a decade away from his final split with the Baptist Church. Lydia's son, T.B., recounts her moment of truth:

> When only 13 she wanted to be baptized and unite with the Baptist Church, but when she was asked to subscribe to the Covenant and Article of the Church, she refused to do so and asked to be baptized on the profession of her faith in Jesus Christ. When the pastor of our home church refused to baptize her, she found another that would and did.[33]

Like her great-great-great-grandfather-in-law (assuming there is such a thing) before the Boston church fathers, and like Campbell in his tricky dealings with the Redstone Baptist Association, Lydia and her convert husband, Barney Knowles, were not left in peace with their beliefs. The Baptist Church in Nova Scotia came after them wielding the heresy ax. They were disfellowshipped from the Baptist Church, an act that left them no alternative but to begin their own church. But, of course, it wasn't their "own church," any more than similar clusters of worshiping excommunicants springing up all over the continent were owned entities. Rather, they were the inductive pieces that would come together in the whole of the Disciples of Christ movement. And they would have no illusions about who owned the church, insisting on the "of Christ" addition to their "Disciples" name to ensure that they never get confused on that point, as had, they believed, their home churches.

By the time T.B. Knowles journeyed to Bethany College in 1867, the year after Alexander Campbell's death there, the Knowleses of Nova Scotia were intimately aware of the great Restoration leader. Upon entering the college grounds across the Ohio River in the newly created, post-war state of West Virginia, Great-Grandpa T.B. was guilty of experiencing a most unRestoration-like feeling of "standing on sacred ground." Later, he would feel similarly awed when he preached in the Bethany church pulpit that had so often hosted Campbell. He would also develop a close son-like relationship with the widowed Mrs. Campbell.[34]

A good number of T.B.'s memorabilia are still carefully kept at Bethany College, but Bethany was and is still a bit short of the Cuyahoga Valley. The link that brought the reluctant family to its Cuyahoga roots was forged by T.B.'s sister, Anne, who in 1873 had her aging parents, Lydia and Barney, come to Akron for their later years. Knowleses would thereafter live within healthy hiking distance of the Crooked River for the next 135 years. But the Nova Scotia connection had one more fascinating card to play. It came in the forms of two men, A.B. Green and Obediah Newcomb, who crossed paths in 1822 at the Baptist church the latter had planted in Wadsworth, just southwest of Akron. Elder Newcomb thereafter spent several uncomfortable years struggling along the familiar path leading away from denominational churches and toward a creed-free, New Testament-based church. He was prodded by young men like Green, whom he baptized on December 28, 1828, and who asked him tough questions about what seemed like clear passages of scripture condemning existing church practices. The increasingly familiar story ended (or began) in February, 1829, when Newcomb led seven other believers, including his two daughters and Green, to establish a Restoration church in Wadsworth. Less than two months later, Lydia and Barney Knowles were married in Nova Scotia. The connection? Actually, there were two. The first is that Newcomb had come to Wadsworth from Nova Scotia. He had left the Canadian Province in 1818, two years *after* Lydia had taken her stand against the Baptists. Lydia must have caused such a unique row in the small province that it is not unreasonable to guess that a man with the strong religious convictions of Newcomb would have heard of her. It is asking too much to completely resist the hunch that Lydia's case may have planted a seed or two in Newcomb's own fertile spiritual garden.

The second half of the link requires no guesswork. A.B. Green, who came to the Disciples under Newcomb's leading, and who spent more hours evangelizing the Cuyahoga Valley for the Disciples than Campbell and Scott combined, made an interesting choice for his later-life mate when he became the second husband of Amanda Baldwin, mother of Cordelia Baldwin. Cordelia was the wife of T.B. Knowles, son of Lydia, the religious renegade of Nova Scotia. (In 1879, they named their second son Almon, which was also the name represented by the "A" in A. B. Green.) The marriage lasted until Green's death in 1887, the year of my Grandpa Errett's birth.[35] Nova Scotia had, without the use of a direct bloodline, radically reduced the need for six degrees of separation in our family linkages.

~~~

From the beginning, Newburgh was something of a Mecca. Small wonder that a great church would one day be planted there. My home church. At first, Newburgh was simply the high ground calling Cleveland's first settlers out of that disease-ridden hole and up to the purer air six miles south. The Kingsburys, the first white family with children on the Western Reserve, deserted Cleveland's real founder, Major Lorenzo Carter, at the mouth of the Cuyahoga at the turn of the nineteenth century. Having barely survived the brutal winter of 1797-98 in Conneaut (one child died of starvation), the Kingsburys could hardly be blamed for not wanting to further risk malarial Cleveland. James Kingsbury opted for starting a mill on the high ground up river that was to become Newburgh. Others, like future Ohio Governor Samuel Huntington, followed, until the heights at Newburgh were looking down on Cleveland in more ways than one.[36]

Spiritual life was not so easily established on the heights. When the circuit riding Joseph Badger visited the five families comprising Newburgh in 1802, he found the higher altitudes decidedly chilly. The not-easily discouraged missionary could find "no apparent piety among any of them," and concluded that "they all seemed to glory in their infidelity."[37] But by 1835, Newburgh had been caught up in the Restoration fever to the extent that the Disciples decided to hold their yearly meeting in that place. It turned out to be a historic event for several reasons, not the least of which was that a church was planted in the revival's churning wake. Like so many other aspects of the frontier, the religious "protracted meeting" is a lost concept today, difficult to envision (unless you can imagine a religious Woodstock) and spiritual galaxies distant from our safe, tame, culturally acclimated megachurches. It usually involved taking over a huge field donated by some faithful farmer. Various stands would be established, each hosting some great preacher doing his thing. An aerial view would have revealed numerous, scattered swarms of humanity, each pulsing to the beat of a particular preacher. To be sure, the Disciples' gatherings were considerably calmer than the Methodist affairs, but the difference was only one of degrees, not kind. The Newburgh meeting of 1835 was held on the farm of Col. Wightman, who personally hosted one hundred of the many hundreds who flocked to his Newburgh farm for four days. The revival spilled over into other surrounding neighborhoods in the evenings, not surprising given the publicity of the event in the Cleveland area. Alexander Campbell was there (he also used the occasion to begin the Disciples' first ministerial training school, setting a precedent for the many bible colleges to follow), as was A.B. Green, who baptized "a large number of converts" on the Monday of the meeting week. Also present were Newburgh's low-lifes, the cynical and sometimes violent scoffers whom the early Restoration preachers tended to refer to as "infidels." This dark side of Newburgh was still well represented in 1835—one suspects that the Disciples chose that site because of *both* adherents and antagonists in the war of faith—one of whom decided to test the disruptive effect of dropping one of Wightman's large trees on a huge tent erected to host one of the preaching sessions. The tree failed to find its mark and Wightman, confronting his guilty neighbor, made a memorable impression by refusing to take vengeance.[38]

The church at Newburgh that would, in later years, be called the Miles Avenue Church of Christ, survived a shaky start. Cleveland historian, William Ganson Rose, bluntly noted that the congregation at Newburgh lapsed within a few years of the great meeting.[39] A. S. Hayden admitted that by April, 1842, the flame at that church had "run low, and its light was nearly extinct." The problem was one facing frontier evangelism everywhere: the absence of permanent ministers. Great meetings and crusades create great excitement, but who is left to nurture the church when the excitement wears off? Not Campbell, not Scott. Fortunately, the languishing congregation at Newburgh was able to persuade Jonas Hartzel to fill that role, which he did by reorganizing the church in April, 1842. Because the church revived and survived, it was able to offer its pulpit to James A. Garfield seventeen years later. Garfield assumed this short-lived ministry in 1859, on the eve of events that would very soon take him to a wider destiny.

Our Sunday ride to the historic Miles Avenue Church of Christ during the 1950s also required a journey to the holy, high ground. Picking up Valley View Road off of Boyden Rd. near the place where the Bacons set up Northfield's first residence, we would follow the straight canal and its twisting Cuyahoga shadow into south Cleveland, the city having long since swallowed Newburgh. Leaving the canal and the river to push on to

Miles Avenue Church of Christ. Cleveland. My childhood church, born in the wake of an early nineteenth century revival in the Western Reserve. Its historic significance paled, in our childhood estimation, to the endless attraction of a dark maze of hallways, superfluous stairwells, and a dungeon-like furnace room through which people had to pass in their finery each Sunday. *Photo from the Walton Hills Church of Christ collection.*

their industrial duties to the north we would turn eastward to make the long climb up a cluttered hill, past the scene of Dad's earlier employment at Mohawk Foundry, past our old house on Beman Avenue, then across the bridge over the railroad tracks immediately west of Broadway. A half-block up Miles Avenue on the right, past Robinson's Drug Store and a large house or two, sat the grand old church, the name of which still brings a nod or two from people scattered across the continent.

The place was a kid's paradise: a jumble of obscure hallways, oddly linked rooms, improbably placed stairways, camouflaged closets, and a host of other testimonies to a time when church architecture was not so burdened with "seeker sensitivity" and visitor comfort. It was a merry tease for young minds endlessly fascinated with "where things came out," especially when they came out in the unlikeliest of places. One of our favorite pathways led through the furnace room, a shortcut used by the adults as well. This was no glorified janitor's closet, nor was the furnace anything like today's sleek little rectangles, efficiently and unobtrusively pumping huge amounts of warmth from a clean burn of natural gas. Rather, the place was less of a room and more of a cavern, the walls of which seemed not so much to end as to slant away into murky corners where perhaps they came together, perhaps not. Decades of hardened coal soot layered much of what was off the main walkway. In the center of the cavern stood the queen herself, belching and bending and creaking, a massive array of pipes and ducts branching oak-like in all directions at once, as if proudly aware that the entire functioning of this swarm of space and activity we called a church rode on her large shoulders.

On the east side of the building were two long stairways running right next to each other, though separated by walls. One began in the basement, the other in the choir preparation room on the first floor. To this day, I wonder why that architect did that—two parallel stairs, each entering the sanctuary a few feet from the other. At the time, I just assumed it was because stairs are more fun that way. I can remember becoming easily and delightfully confused by the two, wondering where I would come out this time. Naturally, the church's strange arterial structure lent itself to some great games of tag and hide-n-seek. I feel sorry for today's kids, who have to make do in box-like sanctuaries with single doors at the front and back leading only in and out, or, increasingly, in theater-like auditoriums fed by wide foyers and boasting airy, symmetrical classrooms. What a dull way to build a church! Occasionally, though, the fun and games exacted a cost. One worshipful Sunday morning, Cousin Jack, leading our chase up the front stairs, barreled into Mrs. Valasic. The collision ruined her nylons, her emotional equilibrium, and our romps—for awhile.

Sometimes, when reports of our rambunctiousness filtered back to Mom and Dad, we would get the "hairy eyeball," that gut-tightening, icy stare that filled the large void between an irritated rebuke and full-scale punishment, and that seldom failed of its objective. (Parents have long since lost the art of the hairy eyeball—I'm sure Mom and Dad were disappointed when it didn't make any of Dr. Spock's books—but with it went a very effective weapon in one of humanity's oldest chess games.) We later learned that Dad had similar troubles during his younger days at this same church. The best of his worst moments came when he decided to take advantage of his position in the sanctuary balcony to communicate with a friend down below on the main floor. He chose, as his means of communication, a paper airplane, the making of which, while not very smart, was clearly more interesting than the only other alternative: listening to the sermon. Had Dad been tried in court for the deed, I'm sure his lawyer would have claimed that he was only responsible for the paper airplane, not the sudden updraft that looped it up over the minister's head and into the choir behind him.

There was really only one moment in my church childhood when I feared that Lloyd, Jack, and I had gone too far. It came one bitterly cold Sunday night when Kenny Kalat stood before the church with his trumpet to tug at our heartstrings with "I Believe." Kenny was probably thirteen or fourteen at the time, maybe four years older than me, a good-natured kid and the son of lifelong family friends, Ruth and Elmer Kalat. His little brother, Keith, perhaps six or seven years Kenny's junior, was the kind of little kid who kept us laughing almost non-stop for years with his impulsive antics. (Keith once turned the end of a worship service into an uproar when he raced into the sanctuary and turned off the lights. Some well-meaning adult had allowed him that important task each week, but on this occasion, Keith had failed to account for the fact that his Sunday School teacher had dismissed class twenty minutes early.) Kenny's immediate problem that night was that the cold ride in from Bedford had partially frozen the valves on his trumpet. Our immediate problem was that we could not stop laughing at the result. In fact, we could not even slow down. You can go through eight straight notes of "I Believe" before having to engage the second valve to change notes, but sooner or later you have to get on with the song, and every time Kenny reached that point, his horn showed symptoms of Bubonic Plague. After each jangling squeak Kenny merely stopped and exercised the valves, showing no more discomfort than a mechanic working on a frozen bolt head, then started over. Each time our giggles became louder, now blasting through the cracks of tightly pressed lips and covering fingers to bounce off reddened necks inches in front of us. Through bleary eyes, I vaguely noticed that Jack was doubled up and actually lying on the sanctuary floor under the pew. Because we were near the front I could feel the weight of hundreds of eyes bearing down on us. Among them, I was certain, were Mom and Dad's, four eyes burning through our backs with what had to be the worst hairy eyeball yet.

Kenny continued to be our worst enemy. If only he had started crying, or run off the platform, or thrown up, (or chosen *any* other song—where was "Deep and Wide" when we needed it?), we probably would have been shocked out of our dangerous silliness. But, after each disaster, Kenny just stood there calmly fingering the valves. The agony dragged on forever—was it two hours?—during which time my mind was beginning to get a damage report through to the rest of my shaking body. The report indicated that my howling laughter was carrying with it a high price tag. What I didn't know was that somewhere back there among what I supposed to be the brooding blackness of adult judgments, my Dad was having a much worse time than us. He, however, had the additional misfortune to be seated in the same pew with Ruth and Elmer and—worse—Keith. Dad said later that he had held up OK through Kenny's first dive, but he knew he was being pushed close to his limit. Still, he might have made it through the second calamity, assuming Ruth was too preoccupied to notice the quivering pew, had not Keith suddenly awakened to what was going on. In a burst of brotherly concern, Keith exploded into his hall-of-fame laugh, a unique combination of throaty cackles and snorts that was usually much funnier than the event that occasioned it. That we did not hear it is a testimony to our own predicament. That Dad could not help but hear it doomed any chance for a socially redeemable ending to the evening. No one could resist Keith's laugh; no one, that is, except a mother whose performing child is in trouble. Dad's laughter fell in somewhere behind Keith's guffawing, and when it did Ruth turned toward Dad and took the concept of the hairy eyeball to a theretofore unimagined height.

Outside the church as it faced Miles Avenue was a small L-shaped concrete slab that might have passed for a driveway except that no one ever parked there, at least not on Sunday. Immediately after the worship service it was crowded with clusters of men seemingly relieved to be out and enjoying a long-anticipated cigarette. They made their escapes even before we kids did, but we knew when we found them there our dads would be good for a dime—two if you had a brother—that we would take on the dead run down to Robby's Drugstore to exchange for brightly colored pictures of Rocky Colovito, Minnie Minoso, and Jim "Mudcat" Grant. And, of course, the slab of pink bubblegum. (Uncle Frank told us that the bubblegum was made out of the horse parts not used for the main product at the glue factory, but that only served to give us momentary pause in our chewing.) It is probably good that the church, whether intended or not, included that chunk of concrete as a haven for the young fathers. In later years, Dad would take key leadership roles in this church, but back then his Sunday morning expression sometimes betrayed a hangdog look. Maybe those men behind their cigarettes and furtive glances were having a hard time reconciling what transpired inside the church with what they had seen overseas. Maybe they had religion "crammed down their throats" as youngsters. Maybe they were a bit embarrassed at playing what, in the 1950s, was still seen as a largely female role: taking the kids to church. Maybe. Mom, who under any circumstances had difficulty getting Dad out of bed in the mornings, used to consider it a victory if we made it to church in time for communion. I guess there was some tension there. But afterwards, out on the concrete slab, everything was all right again. At least, with us.

Some of Dad's discomfort might have come from the somewhat larger-than-life image that the Knowles Family had imprinted on that grand old church for several decades. That relationship probably went back before the turn of the twentieth century: the obituary of Amanda Green notes that she had been living with her daughter, Cordie, and T.B. Knowles in Cleveland for most of the twenty years since A.B.'s death in 1887. I believe that two of Dad's three sisters, Dorothy and Louise, spent some time at the magnificent Miles Avenue pipe organ, and a half-dozen Knowleses (or more) have preached from its pulpit at one time or another. The expectations of anyone carrying the Knowles name were high at this church, and I suspect that Dad squirmed under the weight of it all. Around the time I turned twelve, Mom and Dad decided to leave Miles Avenue (temporarily, as it turned out) in favor of a Disciples church in closer-to-home Macedonia. Grandpa Errett was less than pleased with that move. The conservative "Independents" in the Restoration Movement, among whom were the congregants at Miles Avenue, had just completed a major split with the Disciples of Christ Church, leaving the predictable sour taste in the mouths of the combatants. I recall Grandpa derisively referring to Harold Monroe, Ohio's Secretary of the Disciples, as the "Bishop," deliberately stoking the anti-Catholic fires always burning just below his surface, but perhaps forgetting that Alexander Campbell had often accepted that title as a badge of respect. The bitter rupture was pretty much a fact by 1960, but Grandpa probably would have even disapproved a move to another Independent church, due to his strong Miles Avenue loyalties.

The winds of change create oddly disarrayed landscapes. During the last half of the twentieth century, every other Knowles left the Miles Avenue Church of Christ in one way or another. Mom and Dad alone were left to represent the family name, with the old rambunctious boy and uncomfortable Navy veteran serving many years as an elder well into his 70's. They had returned to the home church after it "moved away from itself:" that is, its namesake. By the time my generation of Knowleses reached maturity, it was apparent that the church would not survive in its rapidly changing neighborhood in Cleveland's southeast side unless some radical changes in attitudes were made. The dwindling number of members did not want to make those changes. At the time, I viewed their thinking critically, suspecting that here was just another case of white-flight from a neighborhood that needed that church as much as we ever did. But I did not have to live with the decisions being made, nor in fact, had I ever faced a similar predicament. Decades later, the last vestiges of my criticism have been long-since swept from my mind. Every church is potentially one generation from extinction, and each is only as strong as its present covering of flesh. The people at Miles Avenue were not social revolutionaries; they were just folks looking for the best ways to worship and, perhaps, thrive once again.

Ironically, the Miles Avenue Church of Christ's move from Miles Avenue in Cleveland followed much of our early route into her urban environment, eventually pushing up the Cuyahoga Valley's eastern ridge to a spot a mile or so above the river just south of Tinker's Creek where Dunham Road turns upward into the steep rise we called Snake Hill. That relocation served as an inducement for Mom and Dad to resurrect the century-old tradition of listing a Knowles family in their old church's directory. In 1985, many of us came back for the church's sesquicentennial celebration. At least a dozen Knowleses were there, covering an eighty-year age span. No one at the ceremony remembered much about the Newburgh Church, but we had enough in the corporate memory bank, including stories that had been passed along to us, to get us a long way back toward 1835.

~~~

The early Western Reserve was the scene of a religious tension as old as the Christian faith itself, and as new as last Sunday. You can sense it just below the surface of the descendants of those first Campbell/Scott/Stone churches, but the issue goes deeper than any single church can take it. In its simplest form, it is an issue of philosophy: Do we know God primarily through knowledge, or through emotions?

In its nineteenth century dressing, the question was whether religion was something you chose or something you "got." "Getting religion" implied almost no active role on the part of the recipient. The idea was that the capricious, almost whimsical Holy Spirit could fall out of the sky on anyone at any moment. It was not so much a case of getting religion as it was getting got by religion, and the sense of mystery inherent in the process proved very appealing to many frontier westerners. Walter Scott's biographer, William Baxter, contemptuously reported witnessing the core of what he saw as the problem in a revival in which a minister was having trouble getting the spark to catch after the penitents had flocked to the mourner's bench. In desperation the minister cried, "O, Lord! here are the sinners desiring to be converted; Lord, they cannot convert themselves; O Lord, we cannot convert them. No one, O Lord, can convert them but thyself;" and then, changing his tone of voice, pleaded: "and now, Lord, why don't you do it?"[40]

A good piece of the issue goes back to the diverse natures of the Disciples' great leaders, Alexander Campbell and Barton W. Stone. Both men had come to the same rational conclusion about the central and singular importance of scripture, and both shared a passion for the unity of Christian believers. But they traveled different roads toward these destinations. The Scot-Irishman Campbell was both brilliant and powerfully educated, having learned early from his erudite father to rely on his mind in his spiritual search. Meanwhile, Stone lacked most of the privileged advantages afforded Campbell. He remained skeptical of Christianity for many of his younger years, dodging religion and Indians as he came to his manhood while Campbell and the Republic were yet mere youths. Perhaps these personal differences decreed that the two men would worship differently—Campbell preferring the controlled environment of the mind, while Stone wandered the wild roads of the soul—but, in any event, they would have been set apart by America's landmark, old-time revival at Cane Ridge, Kentucky, that launched the nation's Second Great Awakening. Campbell was only a boy in Northern Ireland at the time of Cane Ridge, but the twenty-eight-year old Stone was fully capable of relating to his faith what he saw there. And what he saw he deemed miraculous. Although not participating in the charismatic exercises that affected "the devout and unbelieving alike," Stone was convinced that the laughing/singing/jerking/dancing/falling/screaming/barking at Cane Ridge were real manifestations of the Holy Spirit.

> Stone reports that some of the people became amazingly acrobatic, for they would stand in one place and jerk backward and forward with their head almost touching the ground. He says the dancing was heavenly, as if accompanied by angels, and would continue until the subject fell over exhausted. The barking was a description given in derision, for it was actually grunts that accompanied the jerks. Witnesses would see people on hands and knees in the woods, making this noise with uplifted hands, and would report that "they barked up trees like dogs."[41]

The prospect of Thomas and Alexander Campbell finding any kind of comfort zone among such scenes was dubious, at best. Here were two men who turned their own baptism ceremony into a seven-hour event to allow ample time for father Thomas to explain to their large body of followers why they, as sprinkled Presbyterians, were being immersed. (Disciples' historian LeRoy Garrett adds that two young men in the crowd had time to leave the service in order to join another service, that being the American army that was fighting the War of 1812, then return in time to still witness an hour's preaching and the baptisms.) Their faith was nothing if not *understood*. In their world, where knowledge and logic were high trumps, charismatic experiences were off-suited threes and fours.

Perhaps nowhere was the rational nature of the Disciples (Stone's followers were called "Christians.") more in evidence than in their reliance on the debate as a tool for spreading their message. Alexander Campbell's real home was not so much in the pulpit as at the lectern. He once commented bluntly that one good debate was worth an entire year of preaching, a remark that would not find a comfortable roosting spot in even the most unemotional of today's Restoration churches. Had the Disciples' great leaders entered history's stage now rather than the early 1800s, it is doubtful that any of them, other than Walter Scott, would have ever been heard from. Many of today's Christians, like most Americans, fidget their way through fourteen-minute, televised presidential speeches written for ninth grade levels, impatiently anticipating the idiot-proof wrap-up offered by news anchors and commentators speaking in the one-and two-sentence bursts that nevertheless test the attention spans of their listeners. The possibility of herding five thousand moderns into a tent to sit on backless benches for a week to listen to two men debate complex theological issues might find a warmer reception on Mars. This, Alexander Campbell would never have understood.

Nor would James A. Garfield. During 1859, a year in which he was occasionally filling the pulpit of our ancestral home church in Newburgh, Garfield rocketed to the highest levels of Disciples' visibility when he acquitted himself well in a week-long debate against a brilliant and veteran debater, William Denton, in Chagrin Falls. In preparing for the debate, Garfield knew that biblical knowledge would not be enough, that he would also have to meet one of the country's great naturalists on the latter's scientific home turf. Garfield's diary states that he stayed up until midnight every night for weeks prepping for the debate. His biographer sorted through "dozens of pages of headings and quotations arranged in preparation, showing how he [Garfield] reduced to order pretty much all that was known of general science."[42] Garfield himself called the event the "most momentous occasion of my life," and one of his biographers "doubted if any single event in his career had a more important effect."[43]

I have always tended toward the Campbell side of the head-or-heart issue, probably because of a bloodline that carries with it a preference for a rational understanding of things. But scripture is filled with people who, like Rigdon, travailed in the expectation of some very important and impending event. But Rigdon was not a lone wolf in this regard. Barton W. Stone was among the Restoration Movement personalities often leaning away from Campbell-like rationalism and toward Rigdon-like zeal. Leroy Garrett noted a similar trait in Walter Scott.

> Campbell was always positioned, like the North star, unaffected by terrestrial influences; Scott, like the magnetic needle, was often disturbed and trembling on its center, though

always seeking to return to its true direction. While both were endowed with rational powers, Campbell was given more to reason and Scott more to emotion.[44]

I'd like to conclude that I am a north star, and not a shooting star. But that would be too easy. Real faith demands something more.

~~~

"Doyle Thomas is dead," someone said.

"What? How?" I was stung, betrayed by that ancient liar who whispers that teenagers never die.

"Suicide. The police had arrested him and put him in jail. I think he stole a car. He hung himself in his cell."

I continued to stare in disbelief, trying to make these brutal images fit my remembrance of Doyle Thomas. Stolen car. Jail. Suicide. Hanging. Nothing came together. Doyle was just a loud overweight kid whose rounded curves seemed entirely incapable of such sharp edges.

The news bearer, seeing my discomfort, tried to shrug it off for me. "He must have been scared and panicked. I guess he just couldn't face up to it."

Despite the momentary shock, these were not tidings that would cost me sleep or even a meal. I had not seen Doyle for several years, and anyway, we were never friends. In fact, he had no friends that I knew of. Doyle was the kind of kid that everyone liked disliking, one so conveniently irritating and ridiculous as to remove all guilt from the business of childhood cruelty. Mention his name and you could magically smash the icy tension of an argument with a friend, or the embarrassment of a playground failure. Every kid within hearing distance would groan, roll his eyes, and give the knowing smirk of universal agreement about Doyle Thomas.

By almost any standard—then or now—Doyle was a wimp. Obese and imprisoned in horn-rimmed glasses that kept slipping down a perpetually perspiring nose, he was forever worsening his pitiable status by offering expert advice to people who did not want it, and in circumstances where it was least appreciated. I was once the victim of this aggravating trait. It was during third grade, and I had just lost one of the playground fights that helped me to build a 0-2-1 school fight record. (The tie came two years later, when I slugged a pudgy fourth-grader during some interclass bad-blooding at recess. While it was the "fight's" only blow, his younger age and total indifference to the punch forced me and the Olympic judges to call it a draw.) My opponent for one of the two clear losses was Wilbur Siptac, a buddy that I thought was engaging me in a friendly wrestling match. What I had failed to reckon was that since this action was taking place on the high school grounds as we waited to transfer buses, our horseplay might well draw an audience of older kids, one of whom might very well be Wilbur's older brother, Bob. Unfortunately, all of these possibilities bore fruit, which meant, of course, that Wilbur now had a good deal more at stake in this thing than I did. Unaware of this volatile new chemistry, I made the mistake of pinning him. Suddenly my friend became quite vicious in his close-quarter threat to gain vengeance as soon as he could see himself free. I was so surprised at this abandonment of playfulness that I let him up, whereupon he made good his promise with a stiff punch to my stomach. It was a confusing and painful moment, now exacerbated by the jeers of Wilbur's older brother and his friends. I desperately needed an explanation, an apology or, at the very least, some solitude.

What I got was Doyle Thomas.

"You shoulda put a full nelson on him! You coulda beat him with a full nelson. Why didn't you use a full nelson? You shoulda."

My first impulse was to explain to Doyle that I, in fact, did not own any full nelsons, and even if I did, they probably wouldn't have fit Wilbur anyway, since he was taller than me. No matter. My aching midsection and precarious emotional state robbed me of my voice. Doyle wouldn't have listened anyway. I assumed that his advice was given not because I needed to hear it but because he needed to say it. Now, I give some quarter to the possibility that he may have been trying to make a connection in what must have been a very lonely world.

On another occasion, probably earlier, I heard Doyle expounding to someone on the virtues of basketball's hook shot. I snickered knowingly to a friend, and we concluded that there was probably no such thing as a hook shot. How could there be if Doyle Thomas was talking about it?

Later, I would come to learn that Doyle was right about hook shots. I also learned that he was probably right about full nelsons. It is an effective hold, and had I used it on Wilbur Siptac I might still be in the control position out on that playground. Too, I later recalled that Doyle had not saved his advice for the post-fight interview, but had been shouting it at me throughout the tussle. Now that he was dead, my fifteen-year-old mind began to wonder how many other times Doyle might have been right without my knowing it. But it was a question that had been jerked into oblivion at the end of that jail bed sheet.

As a child, it was easy to assume that Doyle was always wrong because we naturally equated fatness, ugliness, and loudness with wrongness. But that equation began to break down for me as the years passed. Pieces of the truth began appearing in the most unlikely places: unheralded books, common events, quiet people. Conversely, some of life's most beautiful packages I found to be empty or, worse, filled with garbage.

God is an odd correspondent. He wraps up valuable messages and treasures in old socks and ripped cartons, then sends them to us. Seldom does he bother to call and tell us that these are on the way, nor does he say much about the appearance of the messenger. I wonder if we throw away most of our messengers without ever realizing that they may be bearing treasures. I wonder if all of us, every kid who ever dismissed Doyle with a sneer or a poke in the ribs, every adult who ever shook her head in disgust and sent him away, helped to throw out Doyle Thomas without so much as a peek inside?

What great secrets were forever entombed in that repulsive frame? What perceptive thoughts, unseen kindnesses, and hopes for the future, however few in number, were yanked from the world's treasure house of potential resources when his large, soft neck snapped? What yet-to-be-formed children, waiting patiently in the recesses of his chemistry for their turns, suddenly saw themselves called forth to expire with their father in a blue spark that instantly short-circuited untold genealogical possibilities?

There is another disturbing image. It is that of God as he receives back his package, as he mulls over the "return to sender" stamp.

The world never hears his lament. "They didn't even open him," he says softly. "They didn't even open him."

~~~

Religion is of ultimate importance only to the extent that it actually deals with ultimately important issues. When it gets bogged down in bureaucracy, personalities, and ego, the Church is no different than businesses, governments or other vested interests. It can easily be much worse. I am increasingly astounded at how much of our physical and emotional lives can be devoted to things that are of no lasting significance, which, in fact, are almost wholly frivolous. Like a colony of unfortunate ants working in the shadow of the gardener who is momentarily going to spade up all of their labors, we continue to doggedly pursue courses beginning and ending in futility, struggling with a twig of sports fanaticism here, a dead wasp of sexual fantasy there, looking no further than our short antennae. Were we to look upward for even the briefest of moments we would immediately see the silliness of the little exercises to which we devote such passionate attention. We, unlike the noble ant, have no good excuse for our spiritual poverty. We make our own choices.

People by the billions have been dying—every one of them, without exception—for untold millennia, yet death still catches us unawares. We feel shocked, betrayed, as if someone has changed the rules in the middle of the game. Few of us are wise enough, disciplined enough, visionary enough to consistently look over the top of the anthill for the real meaning of a real world. In fact, the analogy of the ants insults a species that is at least doing what they were created to do, laboring logically to fill their nests with good things. We fill our ant hills with junk. But, occasionally, someone sees the bright light from above.

Alexander Campbell was such a person, as were many others who permanently affected the way life was and is lived in the valley of the Cuyahoga. It was unthinkable to Campbell that anything could be more important than his relationship with God, so unthinkable that he seldom spent much time defending that particular assumption. Once, however, he

Evening splendor in the Cuyahoga Valley. Forever. *Photo by Tom Jones.*

did so, during his nine-day debate with Utopianist Robert Owen. After Owen had repeatedly returned to stress his unstinted belief in the mechanistic and naturalistic laws of human nature, the Disciples' leader bluntly commented that such arguments were as relevant to a goat as to a man. What about the *real* issues, Campbell challenged in exasperation? "What is man? Whence came he? Whither does he go?"

> Is he mortal, or an immortal being? Is he doomed to spring up like the grass, bloom like a flower, drop his seed into the earth, and die forever? Is there no object of future hope? No God—no heaven—no exalted society to be known or enjoyed? Are all the great illustrious men and women who have lived before we were born, wasted and gone forever? After a few short days are fled, when the enjoyments and toils of life are over; when our relish for social enjoyment, and our desires for returning to the fountain of life are most acute, must we hang our heads and close our eyes in the desolating and appalling prospect of never opening them again, of never tasting the sweets for which a state of discipline and trial has so well fitted us?[45]

These wonderings might have been entertained by anyone or everyone. They would be as natural coming from the mouth of a Chippewa Indian or Ottoman Turk, as from Campbell, for they cut through all human barriers to link with something too important to be entrusted only to race or place or mind; it had to be written on the human soul. "What is man, that thou are mindful of him?" David cried out as he gazed at the heavens and pondered the littleness of earthly kings. It is a question that has echoed off the walls of the Cuyahoga River Valley, but its answer was seldom found in generals or tycoons or mayors or football players. If that answer existed anywhere, it was to be seen in the sad eyes of the mosquito-torn Irishman lugging muck from the canal bottom, or the black man standing in steam ladling silver ribbons of aluminum into a foundry sand mold, or the comprehending squaw of the Indian warrior as she watched her husband and her civilization falling into the swamp waters outside the crumbling walls of their fort. Here were people who probably knew something about life, and those who lived it, and He who gave it.

The problem is that they were largely mute. They took their secrets to the grave. And yet, they are not beyond recall. The same eternal values that they glimpsed through their sufferings drive the Muses to sing their songs. Throughout this journey through two childhoods, I have been teased by the almost heretical impression that I have been bringing once-living things back to life, that the fast-fading wisps of the distant and the dead were being saved, reborn, in a way, before their last specks of light were extinguished by a long darkness. Maybe this is the only way some of them could ever speak. Somehow, the task seems linked to the stuff of ultimate importance. It is worth doing.

There was one more reason why the stories needed telling, and it is this: if life is the most intriguing of this world's wonders, ought we not to say something about it? How can we pass this way without commenting upon it? The vast majority of us are guilty of this omission. We lay quietly in our caskets with folded hands covering a stupendous heart full of joy, grief, anger, hope, and dozens of other life gifts—any one of which is infinitely more fascinating than our most ingenious piece of technology—all unseen, all unheard, all unremembered. Our immediate family will mourn our passing, but it is a momentary thing for even the most loved among us. A few quickly seeded generations later sees us slip into faded photographs in the bottoms of old shoe boxes, then a routine pick-up for some future, faceless trash-haulers. Like sullen children, we offer only grunts to our Father who is pumping us for information about the first day of school. He is dying to know what we thought of it: Did we make any friends? Play any games? Like the teacher? Learn anything at all?

"What *was* it like for you out there today?"

Notes

1. Bloetscher, *Indians*, 97-98.

2. Izant, "David Hudson," 67.

3. (Butler, *Pictoral History*, 4.)

4. Leroy Garrett, *The Stone-Campbell Movement* (Joplin, Mo.: College Press Publishing Company, 1981), 62-64.

5. Hatcher, *The Western Reserve*, 63-64.

6. Bergdorf, *Life Along the Canal*, 4.

7. Hansen, *Westward the Winds*, 49.

8. Florine Morgan, "Appleseed John," in *The Western Reserve Story*, 143-144.

9. Wheeler, "Shakers and Mormons," 95.

10. Mark E. Peterson, "Christ in America" (pamphlet) (Salt Lake City: Corporation of the President of The Church of Jesus Christ of Latter-day Saints, 1982), 2.

11. Joseph Smith, "The Prophet Joseph Smith's Testimony" (pamphlet) (Salt Lake City: Corporation of the President of The Church of Jesus Christ of Latter-day Saints, 1984), 15.

12. *Doctrine and Covenants,* in *The Holy Bible* (Salt Lake City: The Church of Jesus Christ of Latter-day Saints), Sec. 49, v.22.

13. Henry K. Shaw, *Buckeye Disciples* (St. Louis: Ohio Christian Missionary Society, 1952), 84.

14. Hatcher, *The Western Reserve*, 128.

15. Chandler, ed., *Western Reserve Story*, 220.

16. Hatcher, *The Western Reserve*, 124-129.

17. Garrett, *The Stone-Campbell Movement*, 2.

18. Louis Cochran, *The Fool of God* (Joplin, Mo.: College Press Publishing Company, Inc., 1985), 397.

19. Garrett, *The Stone-Campbell Movement*, 245.

20. Cochran, *The Fool of God*, 381-401.

21. Garrett, *The Stone-Campbell Movement*, 381, 221.

22. Shaw, *Buckeye Disciples*, 49.

23. Garrett, *The Stone-Campbell Movement*, 214.

24. Garrett, *The Stone-Campbell Movement*, 386-387.

25. Garrett, *The Stone-Campbell Movement*, 382-386.

26. A.S. Hayden, *A History of the Disciples on the Western Reserve* (n.p., Western Reserve Christian Preachers' Association, n.d.), 209.

27. Perrin, *Summit County*, 577.

28. Shaw, *Buckeye Disciples*, 99.

29. L.V. Bierce, *Historical Reminiscences of Summit County* (Akron: T. & H.G. Canfield, Publishers, 1854), 106.

30. Aliene Miller Williams, "The Knowles Family and Allied Families" (unpublished photocopy, 1932), 38.

31. Williams, "Allied Families," 31-32.

32. Williams, "Allied Families," 43-44.

33. Thomas Benjamin Knowles, "Early History of Thomas Benjamin Knowles" (unpublished photocopy: c 1935), 65.

34. T.B. Knowles, "Early History," 68-69.

35. *Christian Standard*, 12 October, 1907, 38.

36. Ellis, *The Cuyahoga*, 52-57.

37. Harvey Rice, *Sketches of Western Life*, (Boston: Lee and Shepherd, Publishers, 1887).

38. Hayden, *A History of the Disciples*, 404-406.

39. Rose, *The Making of a City*, 141.

40. William Baxter, *Life of Elder Walter Scott* (Nashville: Gospel Advocate Company, n.d.) 19.

41. Garrett, *The Stone-Campbell Movement*, 104-105.

42. Harry James Brown and Frederick D. Williams, eds. *The Diary of James A. Garfield*, Vol.1, 1848-1871, 337. Lansing: Michigan State University Press, 1967.

43. Theodore Clarke Smith, *The Life and Letters of James Abram Garfield*, Vol.1, (Hamden, Conn.: Archon Books, 1968), 125.

44. Garrett, *The Stone-Campbell Movement*, 210.

45. Garrett, *The Stone-Campbell Movement*, 237-238.

Epilogue

And, so, comes the aging process, that ever coursing current that carries us away from our childhoods to other larger waters linked by the shores of memory. But life doesn't end with epilogues, or book publications…or death. Underneath everything there is always an "always" out there, something more, dancing spectrally just beyond our capacity for seeing and fully understanding. On May 31, 2015, Dad passed peacefully away at 90 after a one-day hospital stay and two priceless hours of a fully cogent farewell. It was an exit that could not have been scripted any better. For six years he had been carrying within him an aortic aneurism that could have burst at any moment, as well as kidneys operating at thirty percent efficiency. Just a few weeks before his death, he had been diagnosed with stage four lung cancer. Yet, less than forty-eight hours before he drew his last breath he had been driving around Columbus to meet his regular two-plus hour visit with Mom at the Alzheimer's Center. His breathing was labored, and he could summon only a couple of ounces of energy, as he half consciously refused to allow death to steal the honor he religiously reserved for his wife of seventy-two years.

When my son, Mark, edited this manuscript for me, he surprised me with the observation that the book was more centered on Dad than anyone else except me. That surprised me. I never consciously intended that imbalance; surely Mom was as—and often more—important in my childhood development. But a quick review affirmed that Mark was right. Perhaps I subconsciously yielded to my lifelong admiration for Dad's quietly-borne hardships: a dead mom when he was four, the aftermath of which saw him shipped to a far-away rural home to live with a little-known aunt and uncle for over a year; a childhood lived in the long ravaging years of the Depression; the reality that he was all on his own at fifteen (yet managed to finish high school on schedule); and the co-opting of his remaining youth at eighteen when, within a few short weeks of his marriage to Marie Steadman on December 12, 1942, he was overseas in America's World War II navy for 33 of the next 34 months, during which time he received the news that his first child had been stillborn.

Nevertheless, Mom's background was every bit as hard. So, why the central role for Dad? The son-father thing?

Mom and Dad moved to Columbus in 2008 where they lived with us for almost a year, then bought the house next door. Mom only lasted about a year and a half at home before the extremity of her disease forced our hand to deal with it. Dad said it was the hardest decision of his life. He often said that Mom had taken care of our family for so many years, and now it was his time to take care of her. (He had done so for the preceding ten incredibly tough years.) These past four years he has taken supper with us each night, joined us afterwards for the news and, sometimes, a PBS offering, and shared my daily life more closely than any friend. But none of this would explain my emotional bias in this book; virtually all of it was originally drafted in the 1990s, long before Mom lost her cogency.

It occurs to me now, under the only hours-old miasma of a heavily draped grief, that maybe it had something to do with the fact that I am losing not one father, but two. There is the Dad of these pages, young, strong, overworked, and hoarding his resources for his family's needs. Like most men, the blinders mandated by his focus on duty meant that he did not see much of the world beyond that which demanded his few priorities. He was always a good father but, as this book hints, he sometimes struggled with the heavy load. Then there is the other Dad, the sixty-to-ninety guy who walked paths the vast majority of his demographic peers seldom find or make. Simply put, he grew a new man inside of himself. By the time of his death he was one of the most gracious and, most important, grateful people I have ever encountered.

The accolades for him sometimes strike me from an oblique angle. They come with unexpected strength—"the most humble man I've ever known," most gracious, sweetest, etc.—and from people who swim in larger pools than he did. I'm not the only one taken aback by this. He was flabbergasted and embarrassed when any such compliments came within his increasingly limited powers of hearing. I think that he, like I, was adjusting to the merging of two remarkable lives in his one person.

Meanwhile, teachers run riot in the larger family, including David and Janet, her husband Bob, Jack's wife Gracie, and my wife Lezlee. Lloyd and Jack, both PhD's (history and English, respectively) exceeded 40-year professorial careers at Great Lakes Christian College, in Lansing, and Milligan College in Tennessee (also respectively). I marvel at how many hundreds of lives they have impacted as only fine teachers can. No headlines, there. Tomorrow's suicide bomber will garner more public attention in a moment than Lloyd and Jack have experienced in their combined lifetimes. Yet I have never been more certain of a truth than the one which quietly holds—billions of times daily—that the world turns on the lives of such people, not on the twists of depraved minds.

Dad did not quite make it to see the final publication of this book (though he read the draft three times), but I am intrigued more so than disappointed by the play of time's hand. His death came just a few months before the final production of this labor of a quarter of a century, as if the Creator was painstakingly reminding me that efforts are more important than awards, that journeys eclipse destinations. The Crooked River has known that truth for eons. What matter that her headwaters headed off in the wrong direction for many miles before cupping back to her mouth? Or that an Indian's tiny canoe and a billionaire's massive ore boat the size of three football fields have been equally at ease in her waters? Or that the linchpin of the world's greatest economic forces was driven into the lowest sweep of the river's twisting turns? Or that she would toss up a national park in a boy's backyard? What matters most about the Cuyahoga River and the valley she has carved is not that they lead somewhere, but that they are somewhere.

So, too, for the people along that crooked path, even if it takes a lifetime to realize it is so.

—Jeff Knowles
June, 2015

About the Author

Jeff Knowles was born in Cleveland in 1948 at the front edge of the Baby Boom generation. At age five, his family moved to Northfield—exactly half way between Cleveland and Akron and in the middle of the Cuyahoga River Valley that is at the heart of this book. Despite undergraduate and graduate degrees in history, he did not come to fully appreciate the historic and cultural significance of his home place until middle age. He retired in 2004 after a thirty-year research career in criminal justice, the last twenty-eight of which were served as director of the Ohio Statistical Analysis Center in the Office of Criminal Justice Services.

His written works have included two books, *What of the Night?* (Herald Press) and *Integrity with Two Eyes* (University Press), as well as articles for *OHIO* magazine, *Cincinnati Magazine*, and several others.

Knowles and his wife of forty-six years, Lezlee, have three children: Kathi, Kimberly and Mark, and six grandchildren: Aaron, Andrew, Emma, Kate, Jack and Matthew (pictured here with the author). A diverse group of places have served as his homes, including Cleveland, Northfield, Carter County (TN), Atlanta, and—for the past thirty-nine years, Columbus, Ohio.

Bibliography

Adams, Samuel Hopkins. *Chingo Smith of the Erie Canal.* New York: Random House, Inc., 1958.

Akron Beacon Journal. 13 March 1913.

"Almon B. Green." (obituary) *The Christian Standard*, April 17, 1886, 124.

Bailey, Thomas A. *The American Pageant: A History of the Republic.* Boston: D.C. Heath and Company, 1966.

Barnholth, William I. "George Croghan, Cuyahoga Valley Indian Trader." Northampton, Ohio: the Northampton Historical Society, (undated).

Barry, Dave, "A Guy Who Won't Commit Should Have His Sublimations Examined," Columbus Dispatch, 29 August 1991, 3f.

Baxter, William. Life of Elder Walter Scott. Nashville: Gospel Advocate Company, (undated).

Bergdorf, Randolf S., ed. "Life Along the Canal: The 1849 Journal of Robert Andrew." Peninsula Library and Historical Society, 1990.

Bierce, L.V. *Historical Reminiscences of Summit County.* Akron: T. & H.G. Canfield, Publishers, 1854.

Bliss, Ellen. "Pioneer Women of Northfield." In *Memorial to the Pioneer Women of the Western Reserve*, edited by Gertrude VanRensselaer Wickham. Cleveland: The Woman's Department of the Cleveland Centennial Commission, 1896.

Bloetscher, Virginia Case. *Indians of the Cuyahoga Valley and Vicinity.* Akron: St. Mary's Church, Anglican Catholic, 1987.

Brown, Harry James, and Frederick D. Williams, eds. *The Diary of James A. Garfield.* Vol. 1, 1848–1871. Lansing: Michigan State University Press, 1967.

Butler, Margaret Manor. *A Pictorial History of the Western Reserve: 1796 to 1860.* Cleveland The Early Settlers Association of the Western Reserve, 1963.

Cardinal, Jare R. and Eric J. "Archaeology and History: Some Suggestions from the Historian's Viewpoint In *Ohio's Western Reserve: A Regional Reader*, edited by Harry F. Lupold and Gladys Haddad. Kent: The Kent State University Press, 1988.

Case, Lora. *Hudson of Long Ago*. Hudson, Ohio: The Hudson Library and Historical Society, 1963.

Chandler, Karen. "Valley of God's Pleasure." In *The Western Reserve Story*, edited by Will and Mary Folger and Harry Lupold. Garrettsville, Ohio: The Western Reserve Magazine, 1981.

Cleveland Plain Dealer. 1 April 1991.

Cochran, Louis. *The Fool of God*. Joplin, Mo.: College Press Publishing Company, Inc., 1985.

Condon, George. *Cleveland: The Best Kept Secret*. Garden City, N.J.: Doubleday & Company, Inc., 1967.

Crary, Christopher Gore. "Frontier Living Conditions in Kirtland." In *Ohio's Western Reserve: A Regional Reader*, edited by Harry F. Lupold and Gladys Haddad. Kent: The Kent State University Press, 1988.

"Doctrine and Covenants." In *The Holy Bible*. Salt Lake City: The Church of Jesus Christ of Latter Day Saints.

Eckert, Allan W. *The Frontiersman*. Canada and the U.S.: Little, Brown, & Company, 1967.

Eckert, Allan W. *Gateway to Empire*. Canada and the U.S.: Little, Brown & Company, 1983.

Eckert, Allan W. *Time of Terror*. Dayton: Landfall Press, 1981.

Eiseley, Loren. *All the Strange Hours: The Excavation of a Life*. New York, N.Y.: Charles Scribner's Sons, 1975.

Eiseley, Loren. *The Immense Journey*. New York: Vintage Books, 1957.

Ellis, William Donohue. *The Cuyahoga*. Dayton: Landfall Press Inc., 1985.

Fletcher, Joseph F. *Situation Ethics: The New Morality*. Philadelphia: Westminster Press, 1966.

Garrett, Leroy. *The Stone-Campbell Movement*. Joplin, Mo.: College Press Publishing Company, 1981.

Gieck, Jack. *A Photo Album of Ohio's Canal Era, 1825—1913*. Kent: The Kent State University Press, 1988.

Gooseman, Bessie. *A History of Olde Northfield Township*. Northfield, Ohio: Historical Society of Olde Northfield, 1973.

"Green." (obituary) *The Christian Standard*. 12 October, 1907, 38.

Griffiths, D., Jr. *Two Years in the New Settlements of Ohio*. Ann Arbor: University Microfilms, Inc., 1966.

Grismer, Hiram Karl. *Akron and Summit County*. Akron: Summit County Historical Society, 1952.

"Guideposts for the Current Debate Over Origins." *Christianity Today*, 8 October 1982, 22.

Haley, J.J. *Debates That Made History* (Restoration Reprint Library). Joplin, Mo.: College Press.

Hansen, Ann Natalie. *Westward the Winds*. Columbus: Sign of the Cock, 1974.

Harper, Arthur R. *Ohio in the Making: A Brief Geological History of Ohio*. Columbus: Ohio State University, 1948.

Hatcher, Harlan. "Building the Railroads." In *Ohio's Western Reserve: A Regional Reader*, edited by Harry F. Lupold and Gladys Haddad. Kent: Kent State University Press, 1988.

Hatcher, Harlan. *The Western Reserve*. Revised Edition. Cleveland and New York: The World Publishing Company, 1966.

Hawley, Zerah. "Eastern Criticism of Frontier Religion." In *Ohio's Western Reserve: A Regional Reader*, edited by Harry F. Lupold and Gladys Haddad. Kent: Kent State University Press, 1988.

Hayden, A.S. *A History of the Disciples on the Western Reserve*. N.p. Western Reserve Christian Preachers' Association, n.d.

"Hinckley Awaiting Buzzards," *Columbus Dispatch*. 13 March, 1978, Sec. B, 9.

Horton, John J. *The Jonathan Hale Farm*. Cleveland: The Western Reserve Historical Society, 1961.

Hothem, Lar. "Link With a Lost World." In *The Western Reserve Story*, edited by Will Folger, Mary Folger and Harry Lupold. Garrettsville, Ohio: Western Reserve Magazine, 1981.

Howe, Henry. *Historical Collections of Ohio*. 2 vols. Cincinnati: C.S. Krehbiel & Co., 1888.

Izant, Grace Goulder. "David Hudson: The Howling Wilderness." In *Ohio's Western Reserve: A Regional Reader*, edited by Harry F. Lupold and Gladys Haddad. Kent: The Kent State University Press, 1988.

Jackson, James S. and Margot. *The Colorful Era of the Ohio Canal*. Akron: The Summit County Historical Society, 1981.

Jackson, James S. and Margot. *Cuyahoga Valley Tales*. Peninsula, Ohio: The Cuyahoga Valley Association, 1985.

Jesensky, Joseph D. *An Archaeological Survey of the Cuyahoga River Valley*. Northampton, Ohio: The Northampton Historical Society, Inc., 1979.

Jesensky, Joseph D. *A Tinker's Creek Valley Sketch Book: 1922–1933*. Northampton, Ohio: Northampton Historical Society, 1980.

Ketchem, Richard. "Memory As History." *American Heritage* 42 (November, 1991): 148.

Knowles, Thomas Benjamin. circa 1925. "Early History of Thomas Benjamin Knowles" (unpublished).

Lewis, C.S. *The Magician's Nephew*. New York: Collier books, 1955.

Lindsey, David. *Ohio's Western Reserve: The Story of Its Place Names*. Cleveland: The Press of Western Reserve University and The Western Reserve Historical Society, 1955.

Ludwig, Charles. *Playmates of the Towpath*. Cincinnati: Cincinnati Times Star, 1929.

Lupold, Harry Forrest. *The Forgotten People: The Woodland Erie*. Hicksville, New York: Exposition Press, 1975.

Lupold, Harry Forrest. "Oh! The Joys of Pioneering!" In *The Western Reserve Story*, edited by Will Folger, Mary Folger and Harry Lupold. Garrettsville, Ohio: The Western Reserve Magazine, 1981.

Manchester, William. *The Glory and the Dream*. Boston, Toronto: Little, Brown and Company, 1974.

Map of Northfield Township (circa. 1870), pamphlet file, Northfield historical file folder.

Miller, Grace, Elizabeth Spelman, Kathryn Boyer, and Robert Boyer, eds. *The Story of Independence*. Independence, Ohio: Independence Historical Society, 1979.

Morgan, Florine. "Appleseed John." In *The Western Reserve Story,* edited by Will Folger, Mary Folger and Harry Lupold. Garrettsville, Ohio: The Western Reserve Magazine, 1981.

Ohio Department of Natural Resources and Ohio Agricultural Research and Development Center. 1974. *Soil Survey: Summit County Ohio*. A. Ritchie and J.R. Steiger. Washington, D.C.: U.S. Department of Agriculture.

Parkman, Francis. *France and England in North America: Volume 1*. New York: Literary Classics of the United States, Inc., The Library of America, 1983.

Perrin, William Henry, J. Battle and W. Goodspeed, eds. *History of Medina County and Ohio*. Chicago: Baskin & Battey, Historical Publishers, 1881.

Perrin, William Henry, ed. *History of Summit County*. Chicago: Gaskin & Battey, Historical Publishers, 1881.

Peterson, Mark E. "Christ in America" (pamphlet). Salt Lake City: Corporation of the President of The Church of Jesus Christ of Latter Day Saints, 1982.

Rice, Harvey. *Sketches of Western Life*. Boston: Lee and Shepherd, Publishers, 1887.

Rosenzweig, Roy and Thelen, David. *The Presence of the Past*. New York: Columbia University Press, 1998.

Scheiber, Harry. *The Ohio Canals*. Athens, Ohio: Ohio University Press, 1969.

"School of the Preachers." *Millennial Harbinger*. October, 1835, 478–9.

Shaw, Henry K. *Buckeye Disciples*. St. Louis: Ohio Christian Missionary Society, 1952.

Shriver, Pillip R. "The Beaver Wars and the Destruction of the Erie Nation." In *Ohio's Western Reserve: A Regional Reader*, edited by Harry Lupold and Gladys Haddad. Kent: The Kent State University Press, 1988.

Sinclair, Upton. *The Jungle*. Cutchogue, New York: Buccaneer Books, 1981.

Smith, Theodore Clarke. *The Life and Letters of James Abram Garfield*. Vol 1. Hamden, Conn.: Archon Books, 1968.

Smith, Joseph. "The Prophet Joseph Smith's Testimony" (pamphlet). Salt Lake City: Corporation of the President of The Church of Jesus Christ of Latter Day Saints, 1984.

Tolstoy, Leo Nikolaevich. *The Death of Ivan Ilyich*. In *The Short Novels of Tolstoy*, edited by Philip Rahv, translated by Aylmer Maude. New York: The Dial Press, 1946.

Vince, Thomas L. "Man With a Mission." In *The Western Reserve Story*, edited by Will Folger, Mary Folger and Harry Lupold. Garrettsville, Ohio: The Western Reserve Magazine, 1981.

Wheeler, Robert A. "Shakers and Mormons in the Early Western Reserve: A Contrast in Life Styles." In *Ohio's Western Reserve: A Regional Reader*, edited by Harry Lupold and Gladys Haddad. Kent: The Kent State University Press, 1988.

Whittlesey, Charles. *Ancient Earth Forts of the Cuyahoga Valley, Ohio*. Cleveland: Fairbanks, Benedict & Co., Printers, 1871.

White, George Willard. "The Ground Water Resources of Summit County, Ohio." In *Glacial Geology of Northeastern Ohio*. Columbus: Ohio Department of Natural Resources, 1982.

Wilcox, Frank N. *The Ohio Canals*. Kent: Kent State University Press, 1969.

Wilcox, Frank N. *Ohio Indian Trails*. Cleveland: The Gate Press, 1933.

Williams, Aliene Miller. 1932. "The Knowles Family and Allied Families" (unpublished).

Woods, Terry K. *Twenty Five Miles to Nowhere*. Coshocton, Ohio: Roscoe Village Foundation, 1978.

Worthley, Georgiana. "Down 100 Years With the Old Ohio Canal." *Cleveland Plain Dealer Magazine*, 22 May 1932.

Index

A
Akron, Ohio, xxii, xxiii, 39, 40, 44, 45, 47, 48, 49, 50, 66, 176, 298, 332, 334
American Indians, 6, 146, 183, 185, 291
Atlanta, Georgia, 128–29, 158, 331
Aurora, 176, 177, 178, 292

B
B & O Railroad, 120
Badger, Joseph, 275, 276, 299
Bath, 70, 72, 195, 240
Brandywine Creek, 2, 6, 8, 177, 178, 180, 187, 236, 237, 240, 241, 242, 243, 248, 262
Brandywine Falls, 2, 6, 8, 28, 73, 100, 177, 178, 180, 187, 230, 237, 241, 242, 248
Brandywine settlement, 2, 6, 8, 28, 73, 100, 177, 178, 180, 187, 230, 237, 241, 242, 248

C
Campbell, 285, 286, 287, 288, 289, 290, 292, 293, 294, 296, 297, 298, 311, 313–14, 320
Campbell, Alexander, 288, 292, 297, 300, 308, 310, 312, 318
Canada, 157, 184, 188, 294, 333
canal boats, 44, 47, 57, 58, 125, 126, 162
Champlain, 148, 163
Charlesworth, 245, 246, 247
Cincinnati, 43, 122, 264, 334, 335
Cleveland, xxii, xxiii, 44, 48, 49, 55, 100, 122, 126–27, 142, 240, 261, 270–71, 331, 332
Cleveland Plain Dealer, 58, 333
Coggswell, 195–96
Columbus, Ohio, 60, 66, 120, 122, 123, 152, 154, 255, 261, 263, 326, 327, 331, 334, 336

Connecticut, 230, 231–33, 235, 274
Cuyahoga, xxiv, xxv, 10, 41, 49, 50, 65, 69, 143, 177, 178, 184–85, 236, 270, 279
Cuyahoga River, xxii, 5, 8, 10, 24, 26, 127, 145, 172, 175, 188, 232, 234, 250, 260
Cuyahoga Valley, 28, 29, 65, 67, 145, 146–47, 152, 153, 155, 174, 175, 176, 189, 297, 298
Cuyahoga Valley National Park (CVNP), 2, 45, 68, 72, 128, 185–86, 188, 241, 261
Cuyahoga Valley National Recreation Area, 10, 67
Cuyahoga Valley Scenic Railroad, 118, 128

D
Darwin, Charles, 31, 32
Dayton, 65, 68, 69, 125, 333
Disciples of Christ, 276, 280, 284, 296

E
Eaton, Lee, 90, 91–92, 97, 98
Eckert, 160, 161, 182, 184–85, 249, 333
Eiseley, 33, 34, 35, 333
Eiseley, Loren, 28, 33, 34–35
Ellis, 48, 143, 153, 157, 184–85, 270, 271, 324, 333
Ellis, William D., 68, 69
Erie Canal, 44, 45, 53, 60, 66, 69, 332
Eries, 44, 124, 145–46, 150–57, 162–64, 174–75, 179, 188
Euclid Avenue, 76–77, 117, 174

F
Five Nations, See Iroquois Indian Nation
Flood of 1913, 26, 61-67

G

Garfield, 301, 313
Garrett, 323, 325, 333
Garrettsville, 271, 333, 334, 335, 336
Grand Hunt, 194–95

H

Hale, Jonathan, 70, 75, 77, 99
Hale House, 72, 75, 76, 77
Hatcher, 124, 322, 323, 334
Highland Road, 10, 40, 86, 240, 242
Hinckley, 193, 194, 195, 197, 198
Hiram, 281, 283, 284, 291
Hoggees, 56, 59, 68
Hopewells, 181, 182
Hudson, 77, 235, 236, 248, 275, 292, 293, 333
Hurons, 153, 155, 157, 188

I

Indian Creek, 146, 187
Indian nations, 145, 147, 181
Indians, 6, 144, 145, 146, 147, 148, 149–50, 158, 161–62, 164, 174, 179–85, 189, 234, 249
Indian trails, 45, 173, 176, 177, 179, 234
Iroquois Indian Nation, xxii, 145, 148, 149, 150–51, 153, 154, 155, 156–57, 163, 175, 182, 188–89

K

Kelley, 48, 53, 122–23, 126
Kent, 69, 125, 227, 270, 332, 333, 334, 336
Kent State University Press, 144, 227, 270, 332, 333, 334, 336
Kirtland, 250, 279, 280, 282, 291, 292, 333

L

Lake Erie, xxiii, 41–42, 44, 55, 63, 145, 151, 154, 174, 175, 182, 230, 233, 235, 263
Lake Trail, 174, 175, 176, 177

Little York, 187, 237, 240, 241, 242, 245, 293
locks, xxii, 36, 39, 40, 49, 54, 55, 57, 59, 66, 67, 68, 108, 126
Lupold, 151, 152, 184, 227, 332, 335
Lupold, Harry, 227, 271, 333, 334, 335, 336
Lupold, Harry F., 270, 333, 334

M

Macedonia, 89, 177, 238, 292
Massillon, 47, 48
Medina County, 194, 197, 227, 335
Miles Avenue, 302, 306, 308, 309
Miles Avenue Church of Christ, 133, 301, 302, 308
Mingos, 145, 146, 149, 151, 162
Mormonism, 279, 280, 281, 290, 291
Mormons, 251, 276, 278, 279, 282–85, 289, 290, 291, 292, 336

N

Newburgh, 293, 299–301, 313
New Orleans, 40, 41, 42, 43
New York Central Railroad, 120
Northampton, 68, 117, 177, 185, 332, 335
Northampton Historical Society, 68, 117, 185, 332, 335
Northfield, 40, 86, 87, 90, 91, 177, 179, 237, 238, 240, 242, 246, 247, 292–94, 331
Northfield Center, 89, 177, 178, 238, 240, 293
Northfield Village, 89, 90, 238
Nova Scotia, 294, 297-99

O

Ohio & Erie Canal, xxii, 37, 39, 67, 77, 122–23, 192, 242
Ohio City, 49
Ohio Indian Trails, 144, 173
Ohio River, xxi, 28, 42, 154, 189
Oviatt, William (and descendents), 99, 240

P

Penn Central Railroad, 128
Pontiac, 145, 147, 149, 157, 159, 160, 162
Ponty's Camp, 157, 162, 177, 178, 179
Portage Path, 173, 175, 176, 234
Puritans, 264, 266, 274

R

Railroads, 55, 72, 75, 118-29, 142
Red Lock, 39, 46, 52, 55, 57, 242, 248
Red Lock Hill, 8, 37, 38, 39, 41, 63, 72, 153, 157, 182, 240
Religion, 31, 273-77, 285-87, 307, 310, 317
Rhode Island, 135, 295
Rigdon, Sidney, 284, 287, 299–92, 313

S

Sandusky, 48, 123, 192, 246
Scott, Walter, 287, 290, 292, 296, 298, 301, 310, 312-14
Seneca Indian Nation, 147, 148, 149, 154, 155, 156, 162, 194
Shakers, 276, 278–80, 284,
Smith, Joseph, 279-83, 291
Solon Foundry, 102-04, 106-11, 114, 116, 130
Summit County, 53, 55, 176, 244,

T

Tecumseh, 145, 147, 149, 160–62, 235
townships, 100, 233, 237, 239, 240, 242, 245, 292
towpaths, 53-54, 56

V

Valley Railroad, 65, 126, 127
Viers, Dorsey, 245-48

W

Wallace, Alfred Russel, 31, 32
Water, 22-28, 34, 36, 38, 41, 46, 52, 54, 58, 61, 65-67, 149, 161-62
Western Reserve, xv-xvii, 22, 23, 152, 40, 41, 75, 76, 99, 100, 176, 227, 272-92, 302
whiskey, 42, 43, 52, 195, 249, 250, 275–76
Wilcox, Frank, 173, 176-78, 242
Willow Lake, 177, 178